Dependency under challenge

To Gail, Jonathan, Nicholas and Christopher

Dependency under challenge

The political economy of the Commonwealth Caribbean

edited by
Anthony Payne *and* **Paul Sutton**

WITHDRAWN

Manchester University Press

Copyright © Manchester University Press 1984

Whilst copyright in the volume as a whole is vested in Manchester University Press copyright in the individual chapters belongs to their respective authors and no chapter may be reproduced whole or in part without the express permission in writing of both author and publisher.

Published by
Manchester University Press
Oxford Road, Manchester M13 9PL, UK
51 Washington Street, Dover, N.H., USA

British Library cataloguing in publication data
Dependency under challenge : the political economy
  of the contemporary Commonwealth Caribbean.
  1. Caribbean area – Economic conditions
  2. Commonwealth of Nations – Economic conditions
  3. Caribbean area – Politics and government
  4. Commonwealth of Nations – Politics and government
  I. Payne, Anthony, *1952* –   II. Sutton, Paul
  330.9729′052   HC151

ISBN 0–7190–0970–7

Library of Congress cataloging in publication data
  Includes index.
  Contents: The national level. Jamaica / Anthony Payne – Trinidad and Tobago / Paul Sutton – Guyana / Clive Thomas – [etc.]
  1. Caribbean Area – Economic policy – Addresses, essays, lectures. 2. Caribbean Area – Politics and government – 1945– – Addresses, essays, lectures, 3. Caribbean Area – Foreign economic relations – Addresses, essays, lectures. 4 Caribbean Area – Dependency on foreign countries – Addresses, essays, lectures. I. Payne, Anthony, 1952– . II. Sutton, Paul K.
HC151.D46 1983   337′.09182′1   83–9841
ISBN 0–7190–0970–7

Photoset in Times
by Northern Phototypesetting Co., Bolton
Printed in Great Britain by
Butler & Tanner Ltd, Frome and London

Contents

Contributors	*page*	vii
Preface		viii
Abbreviations		x
Introduction Dependency theory and the Commonwealth Caribbean *Anthony Payne*		1
Part I *The national level*		
1 Jamaica: the 'democratic socialist' experiment of Michael Manley *Anthony Payne*		18
2 Trinidad and Tobago: oil capitalism and the 'presidential power' of Eric Williams *Paul Sutton*		43
3 Guyana: the rise and fall of 'co-operative socialism' *Clive Thomas*		77
4 Grenada: the New Jewel revolution *Tony Thorndike*		105
Part II *The regional level*		
5 Regional industrial programming in CARICOM *Anthony Payne*		131
6 Agricultural co-operation in CARICOM *W. Andrew Axline*		152
Part III *The international level*		
7 Issues in Commonwealth Caribbean–United States relations *Ramesh F. Ramsaran*		179
8 From neo-colonialism to neo-colonialism: Britain and the EEC in the Commonwealth Caribbean *Paul Sutton*		204

9	Commonwealth Caribbean relations with hemispheric middle powers *Vaughan A. Lewis*	238
10	The Commonwealth Caribbean and the New International Economic Order *Denis Benn*	259
	Conclusion Living with dependency in the Commonwealth Caribbean *Paul Sutton*	281
	Index	289

Contributors

Anthony Payne is Senior Lecturer in Politics at Huddersfield Polytechnic
Paul Sutton is Lecturer in Politics at the University of Hull
Clive Y. Thomas is Director of the Institute of Development Studies at the University of Guyana
Tony Thorndike is Head of the Department of International Relations and Politics at North Staffordshire Polytechnic
Andrew W. Axline is Professor of Political Science at the University of Ottawa
Ramesh F. Ramsaran is Senior Lecturer in International Relations at the University of the West Indies in Trinidad
Vaughan A. Lewis is Director General of the Organisation of Eastern Caribbean States
Denis Benn is Chief of the Caribbean Unit of the United Nations Development Programme

Preface

In a telling phrase penned in 1970 V. S. Naipaul characterised the Commonwealth Caribbean as 'the Third World's third world'. He did so to draw attention not so much to poverty in the region, which is by no means as acute as in many other parts of the Third World, but to the sense of hopelessness and despair which pervaded the intellectual atmosphere and conditioned the political will of those elected to govern and develop their countries. Much of this, of course, could be directly attributed to a past of super-exploitation of labour in slavery and indenture, which left the region economically underdeveloped, racially divided, politically subservient and culturally impoverished. But it also referred to a present of utter dependency, itself the legacy of this past and of contemporary material circumstances which saw these countries as too small and too weak to effect meaningful change; condemned, seemingly for ever, to be the periphery's periphery.

This book has one aim above all. It seeks to demonstrate how in the 1970s the states of the Commonwealth Caribbean individually and collectively sought to counter this negative situation by action nationally, regionally and internationally. It evaluates the theory informing the various strategies, the strategies themselves, and their outcomes, both successful and unsuccessful. In so doing it traces the emergence of the Commonwealth Caribbean as a regional sub-system; as a co-ordinated and vocal group in international affairs; and finally as an object of concern, not least to President Reagan who, in viewing developments in the region during the 1980 United States presidential campaign, set aside traditional United States disregard of Central America and the Commonwealth Caribbean in his designation of the area as a 'circle of crisis'.

In assembling the contributors to this volume the editors must thank,

Preface

even if now somewhat remotely, the generous assistance afforded by the University of Manchester and the University of the West Indies, and in particular pay tribute to the exchange scheme initiated between their respective Departments of Government which enabled both of us to benefit from an extended period of research in the Commonwealth Caribbean. This not only gave us the advantage of appreciating at first hand many of the problems surveyed in the book, but also introduced us to a valued and intellectually stimulating circle of friends among whom are counted the various contributors to this work. They all have in common a familiarity and expertise with their subject extending over many years. The opinions expressed are those of the individual authors themselves; the final responsibility for errors and omissions ours alone.

We are grateful to the CARICOM Secretariat for permission to reproduce the map; to David Jessop of the *Caribbean Chronicle* for assistance with photographs; and to Robert Perks for compiling the index. Last, but not least, we would wish to extend our gratitude and appreciation to our wives, Jill and Lorraine, who not only shared with us the comforts and difficulties of fieldwork in the Caribbean but also suffered the trials of bringing up young children at home without our help while we laboured on this volume. To them both we are very thankful.

Anthony Payne
Paul Sutton
January 1983

Abbreviations

ACP	African, Caribbean and Pacific Countries
AMP	Agricultural Marketing Protocol
AOT	Associated Overseas Territory
APC	Agricultural Production Credit Scheme
BTN	Brussels Tariff Nomenclature
CARDATS	Caribbean Agricultural and Rural Development Advisory and Training Service
CARDI	Caribbean Agricultural Research and Development Institute
CARICOM	Caribbean Community and Common Market
CARIFTA	Caribbean Free Trade Association
CBI	Caribbean Basin Initiative
CCC	Committee of Concerned Citizens of Grenada
CDB	Caribbean Development Bank
CFC	Caribbean Food Corporation
CIA	Central Intelligence Agency
CID	Centre for Industrial Development
CIEC	Conference on International Economic Co-operation
CSA	Commonwealth Sugar Agreement
DA	Development Assistance
DAC	Democratic Action Congress of Trinidad and Tobago
DLP	Democratic Labour Party of Trinidad and Tobago
ECCM	Eastern Caribbean Common Market
ECDC	Economic Co-operation among Developing Countries
ECLA	Economic Commission for Latin America
ECU	European Currency Unit
EDF	European Development Fund
EIB	European Investment Bank
ESF	Economic Support Funds
EUA	European Unit of Account
FCH	Feeding, Clothing and Housing Programme
FIC	Farm Improvement Credit Scheme
GDP	Gross Domestic Product
GMS	Guaranteed Market Scheme
GNP	Grenada National Party

Abbreviations

GSP	Generalised Scheme of Preferences
IBA	International Bauxite Association
ILO	International Labour Organisation
IMF	International Monetary Fund
JEWEL	'Joint Endeavour for the Welfare, Education and Liberation of the People' of Grenada
LDC	Less Developed Country within the Caribbean Community
MAP	'Movement for Assemblies of the People' in Grenada
MDC	More Developed Country within the Caribbean Community
MSA	Most Seriously Affected countries
NIEO	New International Economic Order
NJM	New Jewel Movement of Grenada
OAS	Organisation of American States
OCT	Dependent countries associated with the EEC under Part IV of the Treaty of Rome
OECS	Organisation of Eastern Caribbean States
ONR	Organisation for National Reconstruction of Trinidad and Tobago
OPEC	Organisation of Petroleum Exporting Countries
PNC	People's National Congress of Guyana
PNM	People's National Movement of Trinidad and Tobago
PNP	People's National Party of Jamaica
PPP	People's Progressive Party of Guyana
PRA	People's Revolutionary Army of Grenada
PRG	People's Revolutionary Government of Grenada
SDR	Special Drawing Right
SELA	Latin American Economic System
SITC	Standard International Tariff Classification
TNC	Transnational corporation
UNCTAD	United Nations Conference on Trade and Development
UWI	University of the West Indies
USAID	United States Agency for International Development

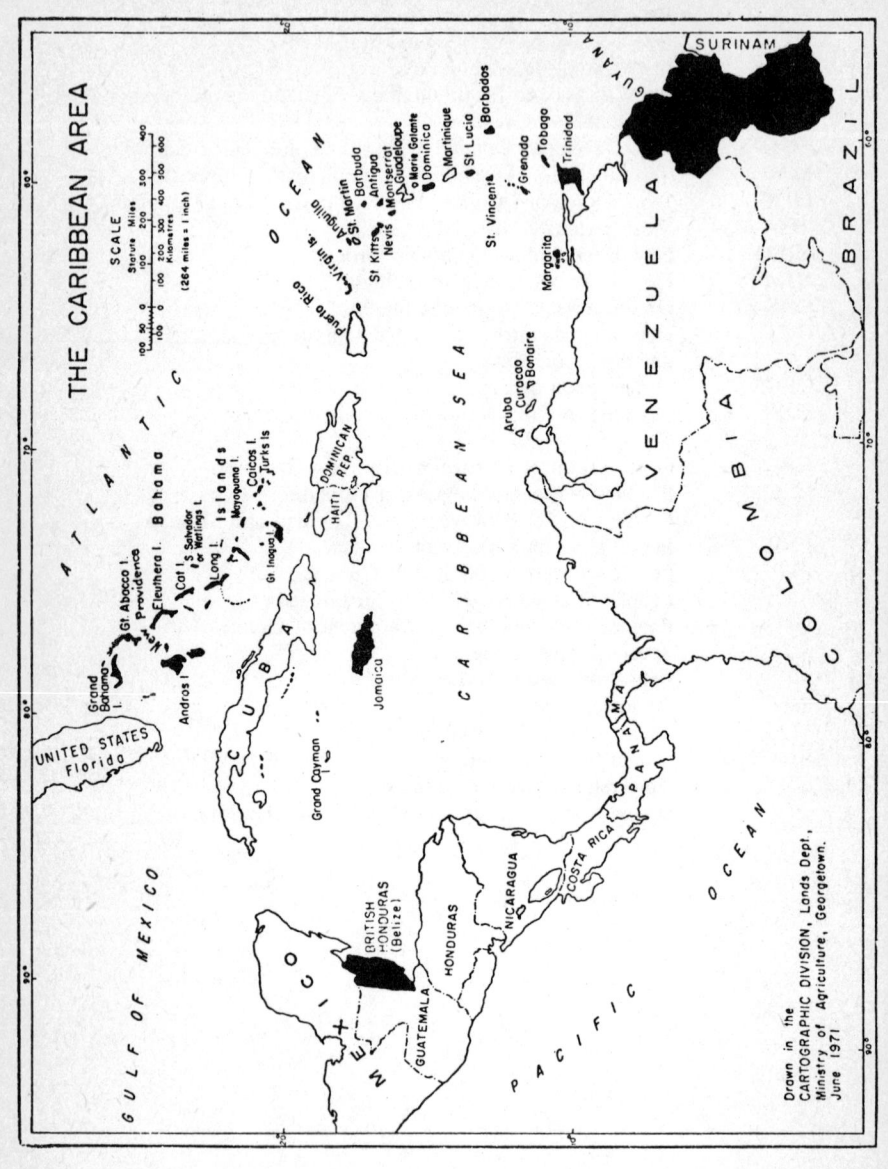

Anthony Payne

Introduction
Dependency theory and the Commonwealth Caribbean

During the last two decades much of the work produced on the political economy of the Commonwealth Caribbean has been conducted within the broad tradition of dependency theory.[1] In this, of course, the Caribbean is not unique, for dependency thinking has come to dominate the study of society, politics and economics in the modern Third World. As is widely recognised, it is an eclectic body of thought, produced by a 'theoretical mingling'[2] of the structuralist approach developed most prominently by Latin American scholars such as Sunkel and Furtado and the neo-Marxist view of underdevelopment popularised most successfully by André Gunder Frank.[3] Its diverse intellectual background has always denied dependency theory the claim to rigour demanded by some theoreticians, but has compensated by permitting a wide range of social scientists, for the most part radical in conviction, to work within a generally understood framework of reference. In this role it has contributed considerably over recent years to developing and extending understanding of Caribbean political economy.

The question of dependence was first discussed in the Commonwealth Caribbean in the early 1960s and quickly came to be seen by a number of economists associated with the New World Group at the University of the West Indies (UWI) as the dominant feature of Caribbean economies. Thus New World Associates[4] in describing the Guyanese economy, Clive Thomas[5] in analysing the monetary and financial arrangements of the Caribbean and Alister McIntyre[6] in assessing the trade policy of the region all felt it necessary to begin by stressing the dependence of the Caribbean economy on the rest of the world: for markets and supplies, transfers of income and capital, banking and financial services, business and technical skills and 'even for ideas about themselves'.[7]

These early insights into the nature and extent of economic dependence in the Caribbean grew into a whole school of thought characterised by what became known as the theory of plantation economy. This theory, developed initially by the Trinidadian Lloyd Best in collaboration with the Canadian economist Kari Levitt, represents the most sophisticated Caribbean variant of the Latin American structuralist view of dependency.[8] The theory consists of an historical and structural analysis of the development of plantation economy in the Caribbean from the seventeenth century to the present day. Best and Levitt emphasised that their primary interest lay in 'isolating the institutional structures and constraints which the contemporary economy has inherited from the plantation legacy', but suggested that the different stages in the evolution of the Caribbean economy should be seen properly as 'successive layers of inherited structures and mechanisms which condition the possibilities of transformation of the present economy'.[9] For the purposes of analysis, they detected three broad phases of historical development: (1) Pure Plantation Economy, covering the period from about 1600 to the abolition of slavery in the British colonial possessions in 1838; (2) Plantation Economy Modified, spanning the hundred years from 1838 to the eve of the second world war; and (3) Plantation Economy Further Modified, which represents the period from 1938 onwards.

The pure plantation economy phase is the crucial one, for it sets the framework for the development of the Caribbean as an 'overseas' economy of a distant industrial metropole. The metropole provided organisation and decision-making, capital, transport, supplies, markets and even transplanted labour from Africa, relegating the Caribbean to the mere locus of production. The local economy was composed more or less entirely of the plantation sector, but had no internal interdependence, each plantation securing its supplies and disposing of its output through its particular metropolitan agent. Conceived as part of the metropolitan economy, where all the benefits of expansion were inevitably felt, pure plantation economy was a highly profitable enterprise in its foundation period; from the point of view of the Caribbean it had entrenched economic dependence.

In the first half of the nineteenth century, according to Best and Levitt, adjustments were forced upon the system, most notably by the abolition of slavery and the removal of imperial preference for sugar, the main plantation staple in the Commonwealth Caribbean. Plantation economy modified is marked by the establishment of a local peasantry

Introduction

on non-plantation land, the recruitment of indentured immigrant workers from India and elsewhere and the rationalisation of the plantation sector, but not by economic transformation. Government continued to be geared towards the maintenance of the plantation sector and thereby checked the expansion of the peasantry and of domestic agriculture.

Further modifications are deemed, finally, to have occurred in plantation economy in the contemporary era, which begins with the second world war. Following the large-scale unemployment, social dislocation and incipient political rebellion which were fostered in the Caribbean in the 1930s by the depression, attempts were made after the war by colonial and ultimately post-colonial governments to bring about greater indigenous economic development in the region. The period of plantation economy further modified has witnessed the institution of schemes of import-substitution industrialisation, the growth of new mineral export sectors like bauxite and petroleum, and the extensive promotion of tourism as an industry, all these activities fuelled by the inflow of foreign private investment. Thus many of the features of pure plantation economy are replicated in this modern period. For Best and Levitt the transnational corporations which have come to play such a dominant role in the Caribbean economy serve just as effectively to integrate the region into the metropolitan economic system as did the joint-stock trading companies of a former era. So far from breaking out of the plantation legacy, the post-war development strategy of the region has only reinforced the constraints of Caribbean economic history.

The other economists of the New World Group readily acknowledge their indebtedness to Best's pioneering work, but they have in their various ways subsequently contributed to the further elaboration of the Caribbean structuralist school. Among them, Brewster has set out perhaps the clearest definition of economic dependence, explaining that:

> Economic dependence may be defined as a lack of capacity to manipulate the operative elements of an economic system. Such a situation is characterised by an absence of inter-dependence between the economic functions of a system. This lack of inter-dependence implies that the system has no internal dynamic which could enable it to function as an independent, autonomous entity.[10]

For his part, Beckford demonstrated more fully what he called the 'underdevelopment biases' of plantation agriculture,[11] Girvan analysed the emergence of mineral-export enclave economies in the Caribbean, with particular reference to the bauxite industry,[12] Jefferson contributed a wide-ranging survey of the post-war Jamaican economy,[13] and others too added elements to the comprehensive new analysis of Caribbean political economy which had flowered by the beginning of the 1970s.[14]

The argument of the New World Group constituted a powerful critique of the condition of economic dependency within the Caribbean, undermining utterly the intellectual credibility of the conventional free-enterprise development strategy pursued by Commonwealth Caribbean governments during the first decade of political independence. Indeed, their vivid excoriation of this strategy as 'industrialisation by invitation'[15] is a memorable aphorism in its own right. Yet their approach was not without its own limitations, which related mainly to its structuralist methodology. In an early discussion of independent thought and intellectual freedom in the Caribbean, Best praised the 'simple rule' of the Latin American structuralists, namely 'to face the reality of what is – of the *particular* situation',[16] and went on to assert the need to understand the Caribbean in terms of Caribbean definitions, thereby escaping, as he saw it, from the irrelevant, rigid and alien formulas of both liberal capitalism and Marxist socialism. This perception led ultimately to what one critic has called 'the basic error of the West Indian New Left',[17] the assumption of Caribbean exceptionalism.

This refers to the emphasis placed by the New World Group on the historical and structural tradition and their consequent tendency to exaggerate the degree of exception enjoyed by the Caribbean from the social and political characteristics of other societies, both developing and developed. As a result, they have virtually ignored the role of class and failed to specify the changing political interests and patterns of collaboration which permitted plantation economy to survive without radical transformation.[18] In the absence of this type of analysis the periodisation of Caribbean history highlighted in the theory portrays only static situations mechanically transposed from one era to another. The approach becomes over-economistic and cannot, therefore, explain why its sensible technocratic proposals for structural economic reform in the Caribbean constantly run into intractable political opposition from vested interests.

However, there is also represented in the analysis of Caribbean political economy the other major strand within dependency theory, the so-called neo-Marxist approach. It developed relatively late, for Marxism has no roots in the Commonwealth Caribbean, either as the basis of intellectual commitment or of political praxis. Marxist political parties have historically been unsuccessful in the region, whilst the study of Marxism had no place within the British colonial education tradition and was slow to take off even within the University of the West Indies, itself by origin an overseas college of the University of London. The Marxist input came second-hand, in good part via the works of Frank, although they were generally translated into English too late to have much impact on the formative period of the plantation economy school. As a consequence, neo-Marxism has never become a prolific form of analysis of the Commonwealth Caribbean. Nevertheless, it has shown an awareness of the major failings of New World and has displayed that wider vision of the internationalisation of capitalism necessary to move beyond such formulations. Its most able exponent has been Clive Thomas, a founder member of the New World Group but quick to move intellectually beyond the structuralist framework. In 1974, after working outside the Caribbean in Tanzania, he published an important book entitled *Dependence and Transformation: The Economics of the Transition to Socialism*.[19] It appeared under the imprint of the Monthly Review Press and drew extensively on the Baran–Sweezy–Frank tradition.

Thomas set his analysis specifically within this line of thought, declaring that a number of studies, by both Marxists and non-Marxists, had established that 'as an observable dynamic the contradiction which has given rise to the reality of the development of underdevelopment in Third World societies is the dialectical process of the internationalization of the capitalist system'.[20] In his view, by the middle of the eighteenth century the economic preconditions of capitalism had spread to such an extent over large parts of the world that Europe's lead in the development of industrial capitalism was not a decisive one; it was made so only by the plunder, slavery and colonial conquest which accompanied European expansion into the periphery. In other words, as Thomas put it, 'European development generated the underdevelopment of the rest of the world by destroying those indigenous social forces which otherwise might have led to the transformation of their precapitalist modes of production'.[21] In their place colonies like those in the Commonwealth Caribbean were offered participation in the global division of labour as providers of raw materials and consumers of manufactures. The long-term consequence was that the productive forces of these countries were detached from their roots in the domestic market and were thus no longer responsive to the needs of the local people.

On the basis of this view of history, Thomas identified what were to him the two most important measures of structural dependence and underdevelopment in the Caribbean as well as other small Third World societies. They were:

> on the one hand, the lack of an organic link, rooted in an indigenous science and technology, between the pattern and growth of domestic resource use and the pattern and growth of domestic demand, and, on the other, the divergence between domestic demand and the needs of the broad mass of the population.[22]

In short, such economies display a pattern of consumption that does not represent the needs of the community and a pattern of production oriented to neither domestic consumption nor domestic needs. As we have seen, this dependence was historically created by act of imperialism but, in Thomas's view, continues to exist in the Caribbean in the modern era despite the inauguration of local manufacturing, the emergence of new export sectors and the impressive national growth rates thereby generated. It is manifest in foreign ownership of key resources, the mode of operation of transnational corporations, the widespread use of inappropriate capital-intensive technology, the parlous state of domestic agriculture and a host of other widely known symptoms of the region's underdevelopment.

Like the other neo-Marxists, Thomas's analysis also has a distinctly political element. He not only describes and explains the historical origins and contemporary economic consequences of the world-wide spread of capitalism, but seeks to understand the political underpinnings of the present conjuncture of production relations and productive forces in the Third World. He draws attention to the failure of an indigenous capitalist class to create its own local material base for self-reproduction in small societies with limited local markets like those of the Caribbean, and to the corresponding existence of dominant social classes who stand to benefit from the continuation of present patterns of underdevelopment. As such, no genuine alternative exists for these countries in the sense that development of their productive forces could take place either by means of the emergence of indigenous capitalism or through socialism. 'That is why', Thomas concluded, 'we have argued time and again that the historical options of these economies are limited either to a comprehensive socialist strategy for transforming the productive forces and liberating the political and social order, or to the continuation of the present neocolonial mode. In the latter case, economic change will continue to be a dependent by-product of developments in the capitalist centre.'[23]

In this very important respect Thomas thus reiterated the essence of the 'stagnation' thesis of Frank and the neo-Marxist development of underdevelopment school as a whole. In the context of the Caribbean his work represented a useful step beyond the structuralists in that it introduced a greater awareness of class and politics to the discussion of dependency and set the whole phenomenon more aptly in the global context of a world system. However, in the last few years the vision of blocked or aborted capitalist development has in its turn come in for close critical scrutiny. The experience of the 1970s, in particular that of the so-called newly industrialising countries, has been taken to show that peripheral capitalism can succeed in some cases and that, as a consequence, the development choices facing Third World countries are not as starkly posed as the former dichotomy between dependency and socialism suggested. Few would go as far as Bill Warren, who has argued that dependence is a myth on the grounds that several Third World countries have achieved high rates of economic growth since the end of the second world war.[24] But a greater realisation undoubtedly exists that dependency and development are not mutually exclusive. Cardoso and Faletto have suggested, for example, that there occurs a kind of 'dependent capitalist development'[25] in the sectors of the Third World integrated into the new forms of monopolistic expansion, by which they mean that the newer forms of dependency associated with the predominance of transnational corporations possess a significant degree of dynamism, rather than stagnation. This is not to say that the social and political costs of this development are not high, nor that it exactly replicates the form taken by capitalist development in the advanced industrialised countries, for neither proposition is sustainable in the light of the work already done by dependency theorists. The argument is really about elevating the debate to a more sophisticated level where tautology is avoided and implicit and unworkable distinctions between dependency and non-dependency are abandoned in favour of a subtler awareness of the varieties of peripheral capitalism that exist in the contemporary Third World.

Unfortunately, Caribbean social science, having contributed to the birth and emergence of dependency analysis, has not been as fully involved in the more recent review of the theory's content and the attempts to go beyond it. The New World Group no longer exists, whilst few of the younger scholars at the University of the West Indies have yet shown the insight and intellectual energy displayed by Best and his colleagues. As a result, the literature on the political economy of the Commonwealth Caribbean does not as yet show much awareness of the main strands of what might almost be called the post-dependency debate about development and underdevelopment. The critique of dependency has been advanced rather more fully in the region at the level of political, rather than intellectual, practice, but generally at the expense of the latter. In respect of political behaviour the dependency school has moved in two different directions roughly consonant with the divergent intellectual approaches represented within the theory.

The New World structuralists have by and large resolved to work for reform within the capitalist system. They have mostly become technocrats, either in the government of their own countries or with international development agencies. Girvan, for example, was head of the National Planning Agency of the Jamaican government under Michael Manley and Jefferson Director of the Central Bank of Jamaica; McIntyre is Deputy Secretary General of UNCTAD and Brewster a senior official of the same organisation. Best, on the other hand, went into politics in Trinidad, became leader of the Tapia party and was centrally involved in the disturbances of 1970 which came close to bringing down the government. However, he subsequently came to support reform of the system, urging a localisation of the Trinidadian economy designed to strengthen capitalism in that country by creating a genuine national bourgeoisie.[26] Recently even he has succumbed to the lure of technocracy by accepting appointment to a United Nations mission to Africa. Virtually every ex-member of the New World Group would thus now see the state as a potential agent of reform within the confines of international capitalism. They are prepared to accept that a measure of development and industrialisation is attainable in the Commonwealth Caribbean. Their changing views have been well described by Courtney Blackman, a former junior member of the New World Group:

> As a young economist in the early 1950's, I was much attracted by their approach and, in fact, made a modest contribution to 'dependence' literature myself in the form of a Ph.D. dissertation on 'Central Banking in a Dependent Economy – the Jamaican experience 1961–7'. To this day I believe that 'dependence theory' provides a useful *descriptive* analysis of under-development in former colonies. Unfortunately, having developed a useful theory of under-development, the scholarship of the New World economists fell apart.

As some of them now admit, they omitted the next logical step – the development of an *operational* model of economic development. The stress is on 'operational'. By operational, I meal 'likely to succeed in real world conditions'.[27]

Blackman is now Governor of the Central Bank of Barbados and, like so many other former New World economists, is working in the 'real world' of government to manage and moderate dependency.

By contrast, the neo-Marxists in the Caribbean dependency school have tended to move more into active political opposition to their respective governments. In doing so they have rediscovered traditional Marxism-Leninism and many – though not Clive Thomas – now espouse 'the non-capitalist path of development', or 'the path of socialist orientation' as it is increasingly being called.[28] The theory posits that the construction of socialism is not dependent upon the prior emergence and full development of capitalism: the task can be begun before the material and productive prerequisites for socialist transition are available and the capitalist phase thereby effectively by-passed or interrupted. The key requirement is a period of socialist orientation, built upon a broad class alliance involving the proletariat, the semi-proletariat masses, the peasants, the revolutionary or democratic strata of the petty bourgeoisie and even the progressive patriotic elements of the emerging national bourgeoisie. The doctrine is now the official philosophy of such left-wing opposition groups as the Workers' Party of Jamaica, led by the UWI lecturer Trevor Munroe, the United People's Movement in St Vincent, led until recently by the former UWI lecturer Ralph Gonsalves, and assorted small radical parties in other parts of the region. It is also endorsed in all essentials by the communist parties of the area, including Cuba,[29] and though not yet fully adopted by the Provisional Revolutionary Government of Grenada has had a profound and increasing influence on several leading members of that government.

The debate about underdevelopment in the Commonwealth Caribbean is thus at present polarised between two extreme positions of co-optation and contestation. Neither is fully articulated in intellectual terms, because of the imperatives of day-to-day action in the respective spheres in which their practitioners are currently engaged and because of due regard to the demands and sensibilities of the different masters they serve. The paradigm of Caribbean dependency has, as it were, dissolved into the very fabric of Commonwealth Caribbean society without changing it in the least, Grenada excepted. This seems an unworthy end in the face of the problems the region still confronts. It stands in urgent need of rediscovery, revival and renewal if the Commonwealth Caribbean's full development potential is to be realised.

Notes

1. For a previous discussion of the emergence of dependency theory in the Caribbean see Norman Girvan, 'The development of dependency economics in the Caribbean and Latin America: review and comparison', *Social and Economic Studies*, 22, 1973, pp. 1–33.
2. Colin Leys, 'Underdevelopment and dependency: critical notes', *Journal of Contemporary Asia*, 7, 1977, p. 98.
3. For general reviews see Philip O'Brien, 'A critique of Latin American theories of dependency' in I. Oxaal, T. Barnett and D. Booth (eds.), *Beyond the Sociology of Development: Economy and Society in Latin America and Africa*, London, 1975, pp. 7–27, and Gabriel Palma, 'Dependency and development: a critical overview' in D. Seers (ed.), *Dependency Theory: A Critical Reassessment*, London, 1981, pp. 20–78.
4. New World Associates, 'The long-term economic, political and cultural programme for Guyana' in N. Girvan and O. Jefferson (eds.), *Readings in the Political Economy of the Caribbean*, Kingston, 1971, pp. 241–65.
5. Clive Thomas, *Monetary and Financial Arrangements in a Dependent Monetary Economy*, Kingston, 1965.
6. Alister McIntyre, 'Some issues of trade policy in the West Indies' in Girvan and Jefferson, *Readings*, pp. 165–83.
7. *Ibid.*, p. 165.
8. See Lloyd Best and Kari Levitt, *Externally Propelled Industrialization and Growth in the Caribbean*, Montreal, 1969, mimeo, and 'Character of Caribbean economy' in George Beckford (ed.), *Caribbean Economy: Dependence and Backwardness*, Kingston, 1975, pp. 34–60.
9. Best and Levitt, *Externally Propelled Industrialization*, Vol. 1, p. 12.
10. Havelock Brewster, 'Economic dependence: a quantitative interpretation', *Social and Economic Studies*, 22, 1973, p. 90. It is worthy of note that this formulation of dependence is remarkably similar to the classic definition offered by the Latin American school, that of Dos Santos in Henry Bernstein (ed.), *Underdevelopment and Development: The Third World Today*, London, 1973, p. 76.
11. George Beckford, *Persistent Poverty: Underdevelopment in Plantation Economies of the Third World*, Oxford, 1972, pp. 183–214.
12. N. Girvan, *The Caribbean Bauxite Industry*, Kingston, 1967, and 'Multinational corporations and dependent underdevelopment in mineral export economies', *Social and Economic Studies*, 19, 1970.
13. Owen Jefferson, *The Post-war Economic Development of Jamaica*, Kingston, 1972.
14. See other articles in Girvan and Jefferson, *Readings*.
15. *Ibid.*, Introduction, p. 2.

Introduction

16 Lloyd Best, 'Independent thought and Caribbean freedom' in *ibid.*, p. 21. Best's emphasis.
17 Ivar Oxaal, 'The dependency economist as grassroots politician in the Caribbean' in Oxaal, Barnett and Booth, *Beyond the Sociology of Development*, p. 46.
18 See Dennis Pantin, 'The plantation economy model and the Caribbean', *Institute of Development Studies Bulletin*, 12, 1980, pp. 17–23, and Marietta Morrissey, 'Imperial designs: a sociology of knowledge study of British and American dominance in the development of Caribbean social science', *Latin American Perspectives*, 3, 4, 1976, pp. 112–4.
19 Clive Thomas, *Dependence and Transformation: The Economics of the Transition to Socialism*, New York, 1974.
20 *Ibid.*, p. 50.
21 *Ibid.*, p. 58.
22 *Ibid.*, p. 59.
23 *Ibid.*, p. 116–17.
24 See Bill Warren, *Imperialism: Pioneer of Capitalism*, London, 1980, especially pp. 185–255.
25 See F. H. Cardoso and E. Faletto, *Dependency and Development in Latin America*, Berkeley, 1979.
26 See Oxaal, 'The dependency economist' in Oxaal, Barnett and Booth, *Beyond the Sociology of Development*, pp. 41–4.
27 Courtney Blackman, 'Speech to the students of Chancellor Hall, University of the West Indies, 14 March 1980', *Caribbean Monthly Bulletin*, 14, 6–7, 1980, p. 45.
28 See Ralph Gonsalves, *The Non-capitalist Path of Development: Africa and the Caribbean*, London, 1981, and, for a critical view, Clive Thomas, ' "The non-capitalist path" as theory and practice of decolonization and socialist transformation', *Latin American Perspectives*, 5, 2, 1978, pp. 10–36.
29 See 'Declaration of the Communist Parties of Latin America and the Caribbean, June 1975' in W. Ratcliff, *Castroism and Communism in Latin America 1959–1976: The Varieties of the Marxist-Leninist Experience*, Washington DC, 1976, Appendix D.

PART I

The national level

The modern nationalist movement in the Commonwealth Caribbean was created out of the depressed economic conditions of the 1930s. Prices for the region's principal export commodities fell, wages were reduced, taxes raised, and unemployment greatly worsened. The crisis of the economy spilled over into politics and gave rise to strikes and widespread popular violence. The first outbreak occurred in St Kitts in May 1935, when unemployed sugar workers occupied estates owned by absentee white proprietors, and from there the wave of protest spread over the next two or three years to Trinidad, Jamaica, St Lucia, St Vincent, British Guiana and Barbados. These events reflected a general political awakening of the West Indian peoples after a century of continuing servitude following the abolition of slavery and arguably could have become a genuinely revolutionary assault upon the derelict edifice of British colonialism in the Caribbean.

In the event, any such potential was swiftly extinguished. The political energy of the disturbances was instead channelled into institutions controlled by an educated middle-class leadership. With official British encouragement, the number and strength of trade unions grew and from this base political organisation was soon extended to the creation of political parties. Before the decade ended there existed, for example, in Jamaica the People's National Party, led by an eminent barrister, Norman Manley, and in Barbados the Barbados Progressive League, led by another lawyer, Grantley Adams. Both parties were committed to the enactment of a wider franchise and the eventual attainment of self-government and both stressed their close relationship with the emerging trade union movement. A new pattern of politics had thus been quickly established in response to the disturbances. It represented a significant advance for the West Indian people but at the same time it effectively contained their protests within political structures with which the colonial power was familiar and could work.

As a result the second world war brought more rapid political than social and economic advance. In the latter field, the Colonial Development and Welfare Organisation made elaborate plans for the economic development of the region, but in the end promised more than it achieved. By contrast, the post-war era saw the enactment of a number of significant reforms in the constitutions of the larger Caribbean territories. Britain's policy was now to guide her Caribbean colonies towards self-government by a careful process of schooling. Tutelage was commenced in Jamaica in 1944, when a new constitution allowing for universal adult suffrage was promulgated. The same advance was granted to Trinidad in 1946 and to Barbados in 1951. In British Guiana not all income and property qualifications were removed till 1952, although here too elections to the Legislative Council had been permitted since 1943.

From these early concessions other constitutional advances followed, but always at a pace dictated by Britain. Every tiny step – adult suffrage, the committee system, the Ministerial system, the transformation of the Executive Council into a more representative Council of Ministers, the establishment of Cabinet government and so on – took on the character of another generous gesture by the benevolent imperial master, rather than a victory on the part of indigenous Caribbean nationalism. The emerging West Indian political leadership was for the most part content to bide its time as gradually the whole complicated apparatus of the Westminster model descended upon the region. In every territory, as it did so, potential successors to the British colonial elite thrust themselves to the fore. Following the precedent of Jamaica and Barbados, new union-party alliances began to appear all over the region, including the smaller Leeward and Windward Islands. New leaders emerged, such as Eric Gairy in Grenada, Robert Bradshaw in St Kitts, Vere Bird in Antigua and Ebenezer Joshua in St Vincent, all of them making their mark in the elections of 1951, the first to be conducted under universal suffrage in these islands.

Slightly different situations developed in Trinidad and British Guiana, the two Commonwealth Caribbean territories whose populations are more or less equally divided between African and East Indian elements. In Trinidad party politics was disorganised and chaotic throughout the 1940s and early 1950s and fell into line only in

The national level

1956, when the People's National Movement established itself as the island's leading nationalist political party under the leadership of Dr Eric Williams. Even then it was exceptional in that it was not reliant on a trade union for electoral and financial support. In British Guiana a dominant party did emerge: the People's Progressive Party, led by the Marxist, Dr Cheddi Jagan. It straddled the racial divisions of the territory and won the 1953 elections, the first to be held under universal suffrage. However, the government it subsequently formed lasted only 133 days before Britain suspended the constitution on the unconvincing grounds that Jagan had embarked upon the creation of a monolithic communist state. That act retarded the country's political development and set it upon the road to racial conflict in the early 1960s.

Nevertheless, apart from this last particular situation, the Commonwealth Caribbean seemed by the mid-1950s to be proceeding calmly towards full self-government and the eventual prospect of political independence. At this point the nationalist movement was diverted into what can be seen in retrospect as the blind alley of federalism. A sense of regional identity and an awareness of the case for regional integration had long been part of the political consciousness of the area. The possibility of 'closer union' between the various British West Indian colonies had been considered over the years by several Royal Commissions, and a variety of schemes of 'functional federalism' were in existence by the end of the second world war. Moreover, the prevailing wisdom of the Colonial Office of the time, accepted by nearly all the new nationalist West Indian politicians, was that the territories of the region were too small to become independent on their own. A federation of all the territories concerned seemed the practical answer and was agreed in principle (excepting British Guiana) at the Montego Bay conference in Jamaica in 1947. It was eleven long years before the structure of the proposed federation had been worked out in detail and the new political unit inaugurated, during which time all its member states had individually moved further along the path of decolonisation. Indeed, the Federation itself was no more advanced: its constitution did not add up to full internal self-government, although that status was granted to Jamaica a year after the Federation was established, and to Trinidad and Barbados before it was conceded to the Federation. In these circumstances the Federation quickly became an irrelevance and a nuisance. The idea that it was an indispensable prelude to the attainment of independence lost all credibility as Commonwealth Caribbean leaders gradually came to perceive that the world-wide process of decolonisation had drifted so far past its original conception of what constituted a feasible new state that it was beginning to encompass territories as small as their own.

Undermined in this fundamental respect, the Federation fell apart in an ignominious fashion and was finally laid to rest in 1962 after just four years of existence. Jamaica had by that stage indicated its preference for independence on its own and proceeded to that status on 6 August 1962. Trinidad briefly toyed with the possibility of forging a new federal arrangement but ultimately fought shy of this and followed Jamaica to independence in the same month. Barbados, in turn, tried to patch together a 'Little Eight' federation of the remaining territories, also changed its mind and became independent in November 1966. Six months earlier, following the reconstruction of its political system around the idea of proportional representation and coalition government, British Guiana (as Guyana) had been given the same sovereign right. A year later the Leeward and Windward Islands, except for Montserrat, were granted so-called Associate Status, an arrangement which conceded full internal self-government but kept responsibility for defence and foreign affairs in the hands of the British government.

With this only a brief paragraph remained to be written in the history of decolonisation in the Commonwealth Caribbean. From the outset Associated Statehood contained within its provisions the right of termination at the instigation of the government concerned. It took some time for the leaders of these very small states to acquire the confidence to cash the independence cheque given to them in the 1967 arrangements. Grenada was the first to break the psychological hold, becoming fully independent in February 1974. Dominica and St Lucia followed suit in November 1978 and February 1979 respectively, and they were joined by St Vincent in October 1979, Antigua in November 1981 and St Kitts-Nevis in September 1983. Thus, apart from a few tiny dependencies, the Commonwealth Caribbean has sought and won the 'political kingdom'.

The merit of that achievement has, of course, been questioned. In the view of some, a mere transfer of power has taken place and 'flag independence' is all that has been achieved. That is a harsh judgement which is unjust to the efforts of the early nationalists of the region. As a priority of development, national units had to be created out of the colonial situation, and they have been. That is not to say that major tasks of social and economic reconstruction do not remain to be undertaken. The execution of that mission has largely fallen to the post-independence leadership of the 1970s and 1980s.

Anthony Payne

1 Jamaica: the 'democratic socialist' experiment of Michael Manley

In the last few years Jamaica has aroused more interest in the eyes of the world than any other country in the Commonwealth Caribbean. It owed this attention to the election of a government publicly committed to the ideology of 'democratic socialism' and determined to reshape Jamaican society, economy and external relations accordingly. The government, formed by members of the People's National Party (PNP) and led by Michael Manley, held office for some eight and a half years between 1972 and 1980 and was witness, in Manley's own estimation, to 'some of the more controversial events'[1] of Jamaica's colourful history. It came to power on a tide of hope buoyed up by the support of Jamaica's underprivileged classes for its wide-ranging programme of reform, but ultimately fell from grace following one of the heaviest electoral defeats ever suffered by a leading political party in modern Jamaica.

The whole Manley experiment thus constitutes dramatic evidence of the problems and possibilities that attach to 'democratic socialist' strategies of reform in trying to overcome dependency in the Third World. This chapter examines the Manley years in Jamaica, assesses the PNP government's particular achievements and failures and sets out some of the lessons that can be learnt from the attempt to bring about radical change in a context of underdevelopment and dependency. It begins by describing the main features of the political economy with which Manley and the PNP had to grapple.

The political economy of Jamaica

At the beginning of the 1970s when the Manley government first came to office the political economy of Jamaica was more complex and diverse than ever before in the island's history. The days of sugar monoculture and the resulting ascendancy of foreign plantation owners were past. Even the prevailing pre-independence pattern of an economy consisting mainly of traders importing finished goods in exchange for the export of primary products (extended beyond sugar to include bananas, coffee, citrus and, most important, bauxite, production of which began in 1952 and grew swiftly to an output of one million tons by 1953 and six million by 1958) had been substantially altered. As in other parts of the region, successive Jamaican governments during the 1950s and 1960s pursued policies of industrialisation aimed at encouraging the establishment of both import-substitution and export-oriented manufacturing enterprises. In addition, tourism was assiduously promoted as a further valuable earner of foreign exchange. By these various means the economy was able to grow throughout the 1960s by an average of nearly 6% per annum.[2]

Despite this apparently creditable achievement, Jamaica retained at the end of the decade a weak and dependent economy suffering from many serious defects. Domestic agriculture had remained stagnant and was the source of continuing poverty in rural areas. Income distribution was more uneven than ever, the share of the poorest 40% of the population in personal earned income declining from 7·2% in 1958 to 5·4% in 1968. Illiteracy, poor housing and unemployment remained the lot of large numbers of Jamaicans. Indeed, the level of unemployment and underemployment had increased hugely, doubling from 12% to 24% during the very period of fast economic growth. The higher wage rates paid in the new mineral and manufacturing sectors encouraged people to forsake low-paid agricultural employment in the hope of finding work in these industries, even though the capital-intensive character of most of the imported technology meant that few jobs were created. The expanding sectors of the economy generally forged very limited linkages with other parts of the economy. A substantial amount of sugar was shipped in a raw state although it was technically and commercially feasible to refine it on the island; the bulk of the bauxite mined was exported as ore despite the advantages to Jamaica of processing it locally; the manufacturing sector consisted largely of 'screwdriver' operations, heavily dependent on imports of raw materials and partly finished components; and the tourist industry was notorious for its failure to create a web of linkages with local agriculture and was thus partly responsible for Jamaica's growing imports of foodstuffs. The effect overall was accurately described as 'a form of perverse growth'.[3]

Much of the explanation lay with the extent and nature of the foreign control to which the Jamaican economy was still subject at the beginning of the 1970s. The island's leading sugar estates were owned by a large British company, whilst the bauxite industry was in the hands of four American and Canadian corporations; many hotels were part of foreign businesses. Banks and insurance companies, a large part of the communications network and even a number of basic public utilities (including the electricity and telephone services) were also in foreign ownership. Only in the manufacturing sector were most firms in majority local ownership, although the distinction between a family firm and the branch plant of a foreign company was exceedingly difficult to draw in practice. As one of the New World economists pointed out, 'a family firm may manufacture a metropolitan product under a franchise the clauses of which are so detailed that the metropolitan enterprise is determining almost all the major managerial decisions – raw and intermediate materials procurement, capital equipment, marketing methods, accounting formats and even employment policy'.[4] What is certain is that economic growth was dependent on private foreign capital transmitted to the island by one mechanism or another but then lost to the local economy by means of profit repatriation and intra-company transfer pricing. Jefferson has calculated that between 1959 and 1969 the funds flowing out of the Jamaican economy actually exceeded all the incoming investments attracted by official incentive policies.[5] By 1972 the bulk of this exchange of capital was with the United States, effectively inserting Jamaica into the domain of US hemispheric power.

The continuing external domination of the economy has had a powerful influence on the structure of the class system. In particular, it has fostered the emergence of a weak local capitalism. The Jamaican capitalist class was initially based on its ownership of land and its control of the colonial distributive trade through import/export agencies and commission houses; in the 1960s it succeeded in adjusting its role in the economy by moving, albeit reluctantly, into local manufacturing. It is very tightly knit, not extending much beyond twenty-one family groupings, and focused upon just five interrelated 'super-groups', the Ashenheims, the Desnoes-Geddes, the Harts, the Henriques and the Matalons, who between them occupy over a third of available directorships in the corporate economy, and enjoys an additional sense of unity derived from the absence of blacks and the predominance of Jews, local whites, Syrians and Chinese in its ranks.[6] The visibility and assertiveness of this class should not, however, be allowed to give an exaggerated impression of its power. As we have seen, it maintained its position in the local economy only by securing subordinate agency relationships with foreign firms and by generally effecting a deferential alliance with foreign capital. It has since become dependent for further growth on state patronage in the form of subsidies, incentives, import protection, tax holidays and the like. Increasingly, therefore, it has been forced to cultivate the local political elite which since independence has come to control the state. As Stone has noted, however, 'the alliance between members or sectors of the capitalist class and ruling politicians is one that grows out of the weakness of that class rather than its strength'.[7]

The unformed character of the capitalist class at the beginning of the 1970s also extended to the working class, the peasantry and the unemployed. The latter typically lived at subsistence level and devoted their energies to the search for individualistic solutions to their dilemma, ranging from crime to 'hustling' and political gangsterism. The urban working class grew in size during the economic expansion of the 1960s, but remained only partially organised. The small size of many of the industrial plants established under the import-substitution programme led to a low level of unionisation and the prevalence of paternalistic labour relations. The unionised sector of the working class did possess a capacity for collective action, but tended to limit its activities to narrowly defined goals such as industrial disputes and wage bargaining. In a similar way, the various agricultural associations which developed to represent the interests of small peasants devoted their efforts mainly to extension work, the improvement of marketing facilities and the distribution of state subsidies. In short, there existed few channels by which ordinary Jamaicans could influence economic policy-making at the national level.

As in other dependent societies in the Third World, the fact of external domination and the underdeveloped nature of local class relations placed heavy demands on the state. It is in a position of economic subordination to powerful overseas interests, but nevertheless can direct the domestic political system. What emerged in Jamaica in the 1950s and 1960s, despite superficial similarities to the political norms of Western liberal democracy, was a clientelist style of politics, whereby the political elite functioned as brokers, channelling state patronage both to the depressed masses and to the capitalist class.[8] The first aspect of this system is easy to understand. The low level of stable employment afforded by the particular pattern of economic growth means that that employment through the state and its various departments, boards and agencies has become crucial as a source of livelihood. A tradition has developed in which political support is exchanged for the material benefits of a job, even a home. The second aspect relates to the ambiguous position of local capitalists within the power structure. They play a significant role in the funding of the political parties which compete for control of the state and they sit in important positions on many of the statutory agencies, like the Industrial Development Corporation, by which the policy of the state is implemented. All of which means that their political influence is considerable. Nevertheless the fact remains that since independence they have been forced to act more and more as subordinate clients of the professional educated elites which dominate both the main political parties. To this extent, the state possesses a degree of autonomy which is at one and the same time both real and qualified, a point nicely made by Stone's sardonic observation that 'the hegemony of international corporate capital presents the only constraint to the petty bourgeois party leadership from consolidating into an independent ruling class'.[9]

Manley's strategy of change

From the moment in 1969 when he became leader of the PNP in succession to his father, Michael Manley demonstrated a clear understanding of the nature of the political economy produced by the post-war era of expansion. As a former trade union leader he was aware that the social benefits of economic growth had been spread thinly and that there were pockets of alienation within the urban environment of Kingston which on more than one occasion during the 1960s had exploded into violence.[10] By addressing himself to this discontent, and adopting a dynamic approach to such issues as unemployment, poverty and political participation, Manley was able to lead the PNP to victory in the 1972 elections. The result reflected the broad range of support which Manley received from the young, the unemployed, large sections of the working class and peasantry, most of the professional and administrative middle class and intelligentsia and even some newer members of the capitalist class disaffected by the economic failures of the previous Jamaica Labour Party government,[11] and could only be interpreted as a decisive mandate for change.

By this time Manley had also developed a coherent vision of the changes he wished to introduce in domestic and foreign policy. In the former sphere, they consisted of three basic commitments. The first was the creation of an economy that would be more independent of foreign control and more responsive to the needs of the majority of the people. This required a wide range of initiatives, including land reform, the creation of co-operative farms, the promotion of worker participation and measures to achieve a more equitable distribution of wealth, but it revolved at heart around the extension of state control into the 'commanding heights' of the economy. In the various public utilities Manley envisaged the swift implementation of complete government ownership but admitted that 'as a matter of common sense and reality, public ownership will have to work together with foreign and private capital'[12] for the foreseeable future in such areas as bauxite, sugar, tourism and banking. He rejected expropriation and asserted that 'the question is not whether to use foreign capital in development planning' but how to bring it into 'harmony with national aspirations'.[13] He also accepted the permanent existence of a mixed economy. In his own words, 'once certain priorities have been overtaken in the field of human resources, infrastructure and certain strategic areas of the economy, private enterprise is the method best suited to the production of all the other goods and services which are necessary to the functioning of an economy'.[14]

The second feature of the strategy was the development of a more egalitarian society not only in terms of opportunity but also in the deeper sense of mutual respect and appreciation. To this end, Manley sought to use the state to ensure that certain basic rights were enjoyed by all Jamaicans. Education facilities, previously narrow and elitist, were to be expanded and opened to all classes; resources in the health service were to be concentrated upon preventive medicine in the countryside rather than sophisticated hospital care in the capital; certain laws were to be amended to improve the legal position of employees and women, especially mothers; and, significantly, the government was to accept responsibility for organising massive work programmes of unemployment relief as a positive social duty. A common theme in all these proposals was a commitment to the use of public expenditure to increase the level of egalitarianism in society.

The third aspect of Manley's programme was political. Manley was critical of the remoteness of traditional multi-party democracy in Jamaica, in particular its tendency to reduce itself to the act of choosing a party to form a government every five years. To this he posed the alternative, not of the single-party state, but what he described as 'the politics of participation'. This called for basic political engineering to 'create the institutions through which people feel continuously involved in the decision-making processes'[15] even as governments faced the need to make compromises or take unpopular decisions. Accordingly Manley proposed the revitalisation of Jamaican local government, the establishment of community councils, the involvement of the public in economic planning, the deployment of the political party as an instrument of mass political education and communication and a host of other measures designed to intensify the level of popular political mobilisation.

The various elements of this strategy of change were conceived by Manley as 'a third path'[16] for Jamaica and the rest of the Caribbean, signifying a rejection of both the Puerto Rican and the Cuban models of development. In his view the former was to all intents and purposes the policy pursued in Jamaica in the 1950s and 1960s, emphasising economic growth and foreign investment but neglecting social welfare. By contrast, the latter was a model of revolution with impressive social achievements but based on the Marxist-Leninist view of democracy which did not allow political rights outside the concept of the dictatorship of the proletariat. For Manley both models also stood condemned by their respective dependence on the two super-powers. It was, as he put it, 'self-evident to us that we want to be pawns neither of East nor West, economically or politically'.[17] Only the third path offered Jamaica and the Caribbean the option of real independence.

This last point is of real importance because it highlights the fact that Manley's programme of change possessed a clearly articulated international dimension. He understood from the outset that domestic reform could be secured only if his government was able to negotiate better terms for Jamaica in all its dealings with the international economy. To some extent this was met by the proposal to bring such industries as bauxite and sugar into partial public ownership, but it was also deemed to require a radical revision of the direction of post-independence foreign policy. Manley worked from the perception that Jamaica was a part of the Third World: he attached considerable weight to the development of a global Third World economic strategy designed to increase collective self-reliance, and he favoured the creation of an organisation of bauxite-exporting countries on the lines of OPEC. More generally, he proposed the adoption of an open foreign policy which envisaged relations with a variety of countries beyond Jamaica's traditional partners, including those with communist ideologies and political systems. Notice was also given of Jamaica's readiness to support wars of liberation on the African continent. In short, Manley sought to 'establish the fact that the entire world is the stage upon which a country, however small, pursues this perception of self-interest'.[18]

Manley's thinking contained, therefore, a number of diverse elements which make it difficult to categorise. In this respect, the ideological label he himself adopted, that of 'democratic socialism', does not necessarily help. Manley has recorded that when the PNP came to choose a name around which to mobilise support for its programme of change it considered three possibilities:

> Christian socialist was rejected on the grounds that it might sound like a political ploy. We decided not to use the word socialist alone because it seemed to invite too much speculation. Quite apart from communism, there were a number of African socialist states organised on a one-party basis. Then again, the local communists were at that time in semi-hiding under the term 'scientific socialist'. Since we were neither communist nor seeking to establish a one-party state, it seemed to invite unnecessary risk to use the term socialist without qualification. In the end, we settled on democratic socialist. The democratic was to be given equal emphasis with the socialist, because we were committed to the maintenance of Jamaica's traditional and constitutional plural democracy; and more importantly, because we intended to do everything in our power to deepen and broaden the democratic process of our party and in the society at large.[19]

One notes the care with which Manley and his party tried to establish the particular nature of the ideas that were to guide their attempt to reshape the political economy, and one can appreciate the reasons for their choice. Respect for democracy was both genuine and the *sine qua non* of securing popular support in Jamaica, where the concept is well embedded in the political culture, whilst there was an obvious attraction to the idea of reaffirming the PNP's original socialist appeal which had lain dormant since the expulsion of a small Marxist faction from the party in 1952. From the analytical point of view, however, 'democratic socialism' is hardly the most precise term in the political vocabulary, the content of the socialism as well as the reality of the democracy begging all sorts of questions quite apart from the matter and mode of their attainment.

It is more accurate to view Manley's approach to politics as a complex, but not inconsistent, combination of several strands of thought. First, he is a nationalist, committed to the service of Jamaica, the assertion of its place in the world and the achievement of a greater degree of economic independence for it within the international capitalist system. Second, he is a populist, hostile to class politics, attracted by the idea of wide mass involvement in decision-making and determined to bring the benefits of reform to all sectors of society. Third, he is a social democrat, holding typically Fabian views on the key questions of social and economic organisation. He learnt his political economy at the London School of Economics in the late 1940s, studying under Harold Laski and admiring the reformist policies of the post-war British Labour government. In sum, all Manley's thinking disavows, either explicitly or by implication, Marxist-Leninist notions of class struggle and proletarian dictatorship. It rests instead on what he has called the 'single touchstone of right and wrong',[20] the notion of equality, which he has made the foundation of his strategy of change for Jamaica in both its internal and external dimensions. 'Social organisation,' he has written, 'exists to serve everybody or it has no moral foundation ... the fact that society cannot function effectively without differentials in rewards together with the fact that men are manifestly not equal in talent must not be allowed to obscure the central purpose of social organisation. That is, and must always be, the promotion of the welfare of every member of the human race.'[21] Armed with this package of ideas and possessed of the power of the state, Manley set about the business of reform.

Reform and re-election: March 1972–December 1976

The first two years of the PNP government brought a number of significant social reforms but no decisive break with the previous direction of economic policy. The reforms included the release of some land to farmers under the Land Lease project, the inauguration of 'crash programmes' of job creation to relieve unemployment, and the introduction of a number of new welfare commitments, most notably the announcement of free secondary education for all. Foreign-owned electricity, telephone and omnibus companies were nationalised, but with compensation, and there was a reorganisation of the various statutory corporations in the state sector which, however, served to ensconce some of the PNP's supporters among the capitalist class in important management positions.[22] In all, it was an even-handed start entirely in keeping with the government's multi-class electoral base.

The year 1974 was a turning point for the Manley administration. The fact of dependency rendered Jamaica more than usually vulnerable to the economic crisis which came to a head all over the world. Its economy simply could not cope with the effects of the sudden rise in the price of oil and the generalised economic recession that followed. In one year, from 1973 to 1974, the island's oil import bill rose from J$65 million to J$177 million.[23] Other import prices rose in consequence, especially food and manufactured goods, putting further pressure on the cost of living and the balance of payments. Foreign capital inflows declined, as did income from tourism, and although the price of sugar reached high levels towards the end of 1974 it plummeted the year after. In addition, the government's own measures of welfare reform had considerably increased state expenditure, which, in an economy lacking the resources to sustain such a sudden expansion, meant that the public sector debt also increased dangerously. Between 1972 and 1974 it rose by 56·7% from J$332·6 million to J$520·8 million; more significantly, the foreign component of the debt went up by even more, from J$117·3 million to J$206·3 million over the same period, an increase of 75·6%.[24] It was the cumulative impact of this severe economic situation rather than political commitment *per se* which forced the Manley government to press ahead with the more radical of its proposed reforms.

The first sign of this change of gear came in January 1974 when the government announced its intention of renegotiating the tax agreements signed with the American and Canadian-owned bauxite and alumina companies. These agreements had not been altered since the early 1950s, when the industry was first set up in Jamaica, and produced only a token tax yield for the government. After four months of inconclusive talks Manley abrogated the agreements and imposed a novel method of raising revenue, a production 'levy' on all bauxite mined or processed in Jamaica set at $7\frac{1}{2}$% of the selling price of the aluminium ingot instead of a tax assessed according to an artificial profit level negotiated between the companies and the government. It was an extremely effective mechanism from the Jamaican point of view, raising the revenue from the industry from a meagre J$22·71 million in 1972 to no less than J$170·34 million two years later.[25] Interestingly, the new policy was supported by, and indeed derived a good deal of impetus from, the Jamaican capitalist class, whose economic well-being, dependent on substantial imports of goods not only for personal consumption but as inputs for their 'final touch' manufacturing industries, was being threatened by a shortage of foreign exchange.[26] Their role was recognised by the appointment of two leading members of the class, Meyer Matalon and Patrick Rousseau, as chairman and vice-chairman respectively of the National Bauxite Commission, which presided over the negotiations with the companies. Many of the benefits of the increased revenue also went to these industrialists. To handle the additional money the government created the so-called Capital Development Fund, whose official purpose was to bolster the productive capacity of the economy. Management was again placed in the hands of Meyer Matalon, who instead proceeded to direct the bulk of its budget into the construction industry, to the particular advantage of companies owned by his fellow capitalists.[27]

As far as the government was concerned the institution of the levy marked only the beginning of a 'bauxite offensive'. It initiated talks with the companies aimed at the purchase of majority control in their local operations, set up the Jamaica Bauxite Institute to provide it with independent data on the technical side of the industry and played the leading part in the formation of the International Bauxite Association (IBA), a producers' cartel inspired by OPEC. The aluminium corporations responded to these moves aggressively at first: they filed a suit with the World Bank's International Centre for the Settlement of Investment Disputes contesting the legality of the levy, pressed the United States government to intervene on their behalf with the Jamaican government and began to transfer bauxite and alumina production from Jamaica to other parts of the world. The Jamaican share of the world bauxite market fell from 19% to less than 14% between 1973 and 1976, whilst that of countries like Australia and Guinea increased markedly.[28] Yet ultimately an accommodation was reached between the companies and the Manley government.[29] The companies realised that new tactics were needed to guarantee continued access to the Jamaican ore for which many of their alumina plants were specifically geared and so resolved to accept the production levy and the government's joint venture proposals. Under these the government purchased all the land the companies owned in Jamaica and 51% of their mining operations, whilst the companies kept their refining plants and retained management control for the next ten years at least. In the unspoken part of the agreements they also maintained their superiority in technology, control of the market and access to capital. In these circumstances the government's challenge to the position of the bauxite companies appears rather more muted.

Perhaps the longer-term significance of the new bauxite policy was that it brought the Manley government to the attention of the United States administration. Bauxite is a mineral of strategic importance and in the mid-1970s the United States drew approximately half its supplies from Jamaica.[30] It seems, however, as if the American government was initially prepared to tolerate Jamaica's actions. Manley has reported that he had an affable meeting with Kissinger on the subject in 1974, in which, he believes, he forestalled any American hostility to the levy.[31] The part of the government's bauxite strategy most likely to have aroused Kissinger's opposition was the formation of the IBA. Potentially, at least, this touched a raw American nerve in that it purported to promote a wider Third World resistance to Western economic interests. Such a commitment was, of course, an integral part of Manley's programme of change and was to contribute to the hostility which he quickly came to arouse within the US government.

The main cause of this was Cuba. Not long after assuming office Manley had incurred the displeasure of the US ambassador, who considered him particularly responsible for the decision of the independent Commonweath Caribbean countries to announce their full diplomatic recognition of Cuba. In September 1973 he attracted further suspicion by flying to the Non-aligned summit meeting in Algiers on board Fidel Castro's private plane. Manley's attendance at the conference indicated the prominence of the role that he and his government intended to play in the Third World movement and, in particular, in the diplomacy attached to the call for the enactment of a 'New International Economic Order'. It also marked the beginning of a friendly relationship between the PNP government and the Cuban regime, which agreed to provide Jamaica with technical assistance in school construction and fishing. In the middle of 1975 Manley paid a state visit to Cuba to give his thanks and at Castro's side made an impassioned anti-imperialist speech before cheering crowds. These moves raised real if unfounded doubts in Washington about the extent of Jamaica's commitment to the West. In American eyes the doubts were more than confirmed by Manley's reaction to the Cuban decision to send an army to assist the Marxist regime in Angola. Manley has described a very different meeting with Kissinger during a short vacation the Secretary of State spent in Jamaica towards the end of 1975. 'Suddenly he raised the question of Angola and said he would appreciate it if Jamaica would at least remain neutral on the subject of the Cuban army presence in Angola. I told him that I could make no promises but would pay the utmost attention to his request.'[32] Kissinger then apparently brought up the separate matter of a Jamaican request for a US$100 million trade credit. 'He said they were looking at it, and let the comment hang in the room for a moment. I had the feeling he was sending me a message.'[33] In Manley's mind the linkage was clear, but nevertheless five days later he publicly announced Jamaica's support for Cuban involvement in Angola. This declaration, made out of a sense of morality typical of Manley's approach to international relations, ensured that US economic aid to Jamaica was embargoed until the end of the Ford administration.

These geopolitical considerations were no doubt the primary cause of the deterioration of US–Jamaican relations, but the situation cannot have been helped by one further important initiative of the Manley government in the aftermath of the economic crisis of 1973–74, namely the PNP's dramatic reaffirmation of its commitment to 'democratic socialism' in September 1974. The launch was widely and deliberately publicised,[34] but the manifesto when it appeared contained little that was new, defining 'democratic socialism' as a commitment to four principles already fundamental to Manley's approach to politics: the democratic political process, the Christian principles of brotherhood and equality, the ideals of equal opportunity and equal rights and a determination to prevent the exploitation of the people.[35] Certainly the direction of the government's policy was not noticeably changed by the declaration of support for socialism. The intention was rather to mobilise the people more actively behind the PNP's strategy of change. Following a loss of popularity in the face of spiralling living costs and persistently high unemployment since he came to power, Manley felt it necessary to revive the populist fervour of his 1972 election campaign. In this sense the reaffirmation of socialism was conceived primarily as a piece of mid-term electioneering.

The campaign did, however, have one important unintended consequence. It marked the limit of the PNP's acceptability to members of the local capitalist class. Until this point they had been content to let Manley deal in the rhetoric of the masses and indulge in any symbolic manipulation that helped to placate the evolving consciousness of the oppressed in Jamaica. They were unhappy about the government's developing friendship with Cuba, yet knew, or thought they knew, what Manley was about. But socialism! That broke all the rules. It seemed to constitute the introduction of ideology into the neutered Jamaican political arena and thus risk the development of a kind of politics that articulated, rather than masked, class interests.

Jamaica: the 'democratic socialist' experiment

From this point onwards, domestic and international opposition to Manley's programme of change came together and worked in tandem. Politically, the local capitalist class turned their energies towards a rescuscitation of the opposition Jamaica Labour Party; economically, they contracted their investments in Jamaica, exported foreign currency illegally and migrated in large numbers to North America. Those who stayed created such a mood of panic in the private sector that the corporate economy was slowly drained of activity. From outside Jamaica there appeared to emanate all the signs of what has become known as 'destabilisation': inaccurate foreign press reports calculated to undermine the tourist industry,[36] unexplained violence and arson in the ghettoes of Kingston, a barrage of vituperation against the Manley government in the columns of the *Daily Gleaner*, owned by the Ashenheim family, and allegations of a strong CIA presence among United States embassy personnel in Jamaica.[37] Most damaging of all was the credit squeeze to which the country was subjected. The US Agency for International Development turned down a request for a food grant, the US Export-Import Bank dropped Jamaica's credit rating from the top to the bottom category and commercial banks ceased all lending to the island. In combination with the decline in the production of bauxite immediately following the imposition of the levy, these measures served to bring about a catastrophic deterioration in the foreign exchange position.

In response Manley allowed the process of socialist mobilisation to take off more intensively than the terms of his original strategy dictated. Whilst the economy was temporarily shored up by means of bilateral loans from friendly governments, such as those of Trinidad and Canada, the attack on the country by imperialist forces was made the theme of the PNP's 1976 election campaign. Until then Manley had endeavoured to preserve a balance between the left and the right in the party, but now he gave his support to the left and allowed the party secretariat under the leadership of Dr D. K. Duncan to organise a heavily ideological campaign. Even the communist Worker's Liberation League offered him 'critical support'. In a sense it was hugely successful, because, despite the desperate condition of the economy, the PNP was re-elected in December 1976 with a massive majority, winning forty-seven out of the sixty seats. In fact, although nobody realised it at the time, the victory represented the failure of Manley's initial conception of the politics of change. The PNP no longer represented the majority of people of all classes. Its support among the capitalists and the middle class had disappeared and been displaced by gains among the working class and the unemployed.[38] The overall effect was to polarise class voting patterns and rob Manley of his chance of leading a genuinely populist movement of reform.

The IMF and retreat: January 1977–October 1980

In the heady atmosphere that followed the election the extent to which the nature of Manley's victory had undermined his original strategy of change was not realised. Instead the result was interpreted by many in the PNP as an endorsement of the socialist mobilisation of the preceding year. Before the election Jamaica's dire economic situation had been thought by the government to necessitate an appeal for assistance from the International Monetary Fund (IMF). A provisional set of measures had been worked out and the foreign exchange markets closed as a prelude to their implementation. After the election, however, this policy was cast aside in the face of intense opposition from the party's left wing. They were doubtful of the efficacy of the monetarist approach of the IMF when applied to Third World economies and argued that, in the short run, Jamaica could survive the foreign exchange crisis by careful rationing of earnings, supplemented by loans and other kinds of material support from socialist bloc and progressive OPEC countries. Manley's personal position is not known, but he articulated the government's defiance of the IMF powerfully in a speech to the nation on 5 January 1977:

> the International Monetary Fund, which is the central lending agency for the international capitalist system, has a history of laying down conditions for countries seeking loans ... this government, on behalf of our people, will not accept anybody anywhere in the world telling us what to do in our country. We are the masters in our house and in our house there shall be no other master but ourselves. Above all, we are not for sale.[39]

Accordingly, the government announced a very different programme of economic measures to the deflationary package implicit in the proposed IMF agreement. To be sure, certain moves were perhaps intended to keep that latter option open, such as the imposition of a pay moratorium and a higher tax on petrol, but they could not disguise the dominant influence of the left on the government's post-election thinking. There was to be no devaluation, which was the main demand of the IMF. Several left-wing economists from the University of the

West Indies under the leadership of Norman Girvan were brought in to strengthen the national planning apparatus. Radio Jamaica was to be purchased as part of a political education programme led by the newly formed Ministry of National Mobilisation, whose first head was D. K. Duncan, the PNP's left-wing General Secretary. Negotiations were also announced for the take-over of three foreign banks, including Barclays, and the island's only cement factory, owned by the Ashenheim family. Finally, preparation of a Production Plan was set in motion, to be based on suggestions made by the ordinary people. The strategy as a whole envisaged the mobilisation of Jamaica's exploited classes against local and foreign capitalist control.

It turned out to be only a brief interlude in the political economy of the Manley years. As Fitzroy Ambursley has observed, these initiatives brought the PNP to 'the Rubicon of retreat or social revolution'.[40] At this point Manley could have embarked fully upon the committed socialism of class confrontation, a road he had tentatively set out upon, albeit under pressure of external events, during the mobilisation of 1976 and in the immediate aftermath of the election. The goal of a populist programme of change supported by all classes in society would have had to be abandoned and a different sort of politics erected in its place. There can be no doubt that this was the aim of the left, both inside and outside the PNP. There were, however, still those in the party who shied away from this prospect and would, as Manley put it, 'have looked askance at all these young Turks of the left, often sporting beards, tams and jeans, playing so prominent a part in affairs'.[41] The left themselves argue that in the final analysis Manley himself was to be found in this category, betrayed by 'his lack of confidence in the capacity of the masses of black Jamaican people to assert their productive creativity', a view that 'derives from a brown Jamaican petit-bourgeois perspective'.[42] Certainly nothing in the evolution of Manley's political thought would have predisposed him to accept the ideology of class politics being urged upon him. As noted earlier, his thinking had developed along entirely different lines. At any rate, he retreated. The announcement that negotiations would resume with the IMF was made in April 1977, from which moment onwards the story of the Manley government becomes, to all intents and purposes, the story of its relationship with the Fund.

Manley has subsequently claimed that he had 'no choice at that time'[43] but to do as he did. Within the limits of his own politics, that may well be right. At home, the call for 'people's planning' produced a massive response, but of half-baked, unrelated and often trivial ideas, and ultimately foundered on the harsh reality of the lack of technical and economic knowledge at the mass level. Nor was there ever very wide support for the anti-imperialist implications of the left's strategy, a national poll conducted by Carl Stone in May 1977 showing, by contrast, that 76% of the population favoured the receipt of US aid[44] – hardly the ideal political base from which to challenge the hegemony of international capitalism. Abroad, the constraints were even tighter and, from the left's point of view, not short of irony. As part of the post-election plan tentative efforts were made to seek financial aid from the socialist world, including an unpublicised approach to the Soviet Union. They proved unsuccessful:[45] Jamaica had not prepared the ground in any way for forging the connection and was not otherwise of sufficient political or military significance to attract large-scale Soviet support. At the same time as this rejection was becoming manifest, signals were emanating from Washington that the new Carter administration, which had replaced the Ford–Kissinger team at the beginning of 1977, was desirous of a more friendly relationship with the Caribbean, beginning with Jamaica. Prompted by Carter's ambassador to the United Nations, Andrew Young, an old friend of Manley, the aim of American policy seemed to be to coax Jamaica back into the Western camp. The opening was seized upon by the moderates in the Manley government, who included P. J. Patterson, the Foreign Minister, to argue for a return to the traditional focus of foreign relations. The conjuncture of these two pieces of diplomacy was critical and, in so far as the United States made it known that the resumption of aid would be easier if Jamaica repaired its relations with the IMF, paved the way for the decision to resume talks.

Initially Jamaica was able to wield a good deal of bargaining power in its negotiations with officials of the Fund.[46] In the early rounds the main sticking point was the scale of the devaluation required. The government proposed a dual exchange rate: a 37·5% devalued 'special rate' applicable to traditional exports and inessential imports, but a 'basic rate' at the old level for government transactions, essential imports of food and medicines and, most important, bauxite exports, the value of which would thereby be maintained. The Fund opposed this and the government's slight easing of its wage guidelines, but Manley was able to utilise his personal friendship with Prime Ministers Callaghan of Britain and Trudeau of Canada to bring some leverage to bear on the board of the IMF. In the end, however, it was probably the readiness of the Carter administration to 'lean' on the IMF on Jamaica's behalf that forced the Fund to settle largely on Jamaica's terms. The amount to be provided was US$74·6 million over the two years to June 1979, subject to the economy meeting the well known IMF performance tests. In the circumstances the agreement was relatively satisfactory from Manley's point of view. Additional foreign exchange would become available, the basic imported goods on which the island's poor depended would be spared a huge increase in price, and the government's general programme of extending public ownership could even be slowly advanced. In 1977 the National Sugar Company took over several more of the island's sugar estates and a State Trading Corporation was established with a view to eventually handling all imports.

However, the embrace of the IMF was not to prove so tolerant for long. In a development utterly unexpected by Manley, at least, the economy failed the first performance test in December 1977. On the day in question the net domestic assets of the Bank of Jamaica exceeded the required ceiling of J$355 million by a mere J$9 million, or 2·6%. The Fund seized on this as evidence of irresponsible domestic economic management, suspended the next tranche of the loan and returned for further talks to be conducted in an altogether harsher manner. There is no need to follow every aspect of the negotiations, for there was little that the government could do but agree to a virtually total overturn of its previous economic policies. A new agreement was finally concluded in May 1978, providing US$240 million over three years, but depending upon the introduction of, in Manley's words, 'one of the most savage packages ever imposed on any client government by the IMF'.[47] It contained as its centrepiece the reunification of the exchange rate, along with a further immediate general devaluation of 15%, to be followed by a 'crawling peg' arrangement under which there would be further mini-devaluations every two months (which amounted to 15% over the first year of the programme). There were also to be drastic cuts in the size of the public sector, a cutback in the operations of the State Trading Corporation, sharp tax increases, price liberalisation to guarantee a 20% rate of return on capital invested by the private sector, and finally a mandatory limit on wage settlements of 15% a year, in comparison with the predicted increase in the cost of living of some 40%.

By this stage, what else could the government do? In terms of domestic politics, the PNP left were a defeated force in the struggle within the party, their decline symbolised by Duncan's resignation as

Minister of National Mobilisation in September 1977, whilst the moderates and conservatives who dominated the Cabinet could neither prevail on the IMF to soften its position nor think of any alternative source of foreign exchange. As the tone of his speeches and public appearances at the beginning of 1978 indicated, Manley himself was subdued. Internationally, the Jamaican government had lost even the qualified support of the Carter administration, partly as a result of internal changes within that government, notably the declining influence and eventual resignation of Young, but also because Manley insisted throughout Jamaica's economic traumas on maintaining a radical and pro-Cuban foreign policy. He entertained Castro on a state visit to Jamaica in October 1977 and repeatedly praised the social and economic achievements of the Cuban revolution. He also continued to assert aggressively Jamaica's commitment to a New International Economic Order. In the light of his experience of the IMF one can appreciate his reasons for doing so, but in the short term – and that was by 1978 the only consideration relevant to the Jamaican economy – it was not the most effective way of improving his country's bargaining position with Fund officials.

For Manley the tragedy of all this is that the IMF recipe could not work for a structurally dependent economy like Jamaica's, so lacking in productive capacity. It is now a commonplace of political economy analysis to say this,[48] in good part as a result of Jamaica's experience. The fact is that although the Manley government proceeded to carry out every single aspect of the new agreement, in both letter and spirit, the economy never recovered. The harsh economic medicine did not generate additional inflows of foreign capital or halt the outflow of Jamaican capital, with the result that the economy became locked into a vicious deflationary spiral, with predictable consequences for employment and living standards. Even the bauxite levy was lowered slightly in an attempt to induce the companies to increase production and expand their investment in the industry. Yet in December 1979 the economy once more failed a performance test, again largely because of factors outside governmental control, in particular further increases in the price of oil. The Fund demanded huge public expenditure cuts as the price of continuing the assistance programme, only to find that opinion had hardened in the PNP and in the government during the course of the long and difficult relationship with the IMF. In a clear sign of the changing mood, Dr Duncan had been re-elected to the post of General Secretary in September 1979 and had set in train a detailed discussion

Jamaica: the 'democratic socialist' experiment

within the party of the government's economic policies. The impact was catalytic: a special delegates' conference in January 1980 gave enthusiastic support to proposals that the government should find a 'non-IMF' path and urged it to resist the full extent of the budget cuts being demanded. Eventually, in March 1980, the National Executive Committee of the party voted by a two-to-one majority to break with the IMF and start implementing an alternative programme. The Cabinet accepted this, and Manley announced that elections would be held before the end of the year to enable the people to decide what economic strategy the country should follow.

This was to be the final lurch in what has been referred to as 'the zig-zag politics'[49] of the Manley regime. A new left-wing Finance Minister was appointed and some attempt made to renegotiate the country's commercial debts, but little emerged by way of concrete policy.[50] Once his original strategy had been upset, paradoxically, by the great electoral success of 1976, Manley was never again able to develop a coherent approach to the management of the political economy. The pattern was indeed one of endless vacillation in which he danced uncomfortably in turn with the domestic left and the international right. As he did so the economy deteriorated and destroyed whatever chance he and his party had of being re-elected. In elections finally held on 31 October 1980 the PNP were massively defeated by the conservative Jamaica Labour Party, winning only nine out of sixty seats in Parliament, the party's lowest-ever representation.[51] In the view of Norman Girvan, the UWI economist and head of the National Planning Agency under Manley, the IMF had at last 'succeeded in manipulating the Jamaican political process to re-establish the political conditions in which the previous model of dependent, unequal capitalist development may function'.[52]

Conclusion

The heavy electoral defeat suffered by the PNP was hardly surprising in the light of the overall record of the Manley government over the preceding eight and a half years. It had promised so much and in the end achieved so little. The economic statistics alone are revealing: between 1974 and 1980 there was a cumulative fall of 16% in gross domestic product, whilst unemployment rose from 24% to 31% and the cost of living by an enormous 320%. The exchange rate against the US dollar moved from approximately J$0·88 to J$1·76 and, as we have seen, foreign exchange reserves and foreign capital inflows fell dramatically over the period.[53] One could continue the catalogue of figures, but it would not brighten the picture. As far as ordinary Jamaicans were concerned, the reforms of the Manley government had produced a severe decline in living standards, lower levels of employment, acute shortages of basic goods in the shops and a mood of depression pervading the whole economy and society. Against this dismal background the government's few achievements in the social field and in foreign affairs cannot be said to count for much.

What went wrong? Manley's own analysis has emphasised the question of 'destabilisation',[54] instigated by the United States government, implemented ultimately by the IMF and abetted by treacherous opposition forces in Jamaica. It is part of the explanation, of course, but it does not say everything and too easily becomes a misleading shorthand expression of the fact that the world which Manley took on was often hostile to his aims. Obviously international factors bore heavily, and at times oppressively, upon his regime, but their impact can be properly understood only in relation to the type of politics which he was trying to pursue.

To go back to the beginning, Manley's initial strategy of change was sensibly conceived. It was not likely to be easy to implement, but there was no reason to think it impossible either. The post-colonial state enjoyed a significant degree of autonomy and, if led with sufficient political skill and sense of *Realpolitik*, ought to have been able to restructure the political economy along the lines Manley proposed without arousing the implacable opposition of either the indigenous capitalist class or international capitalist interests as a whole. The former had a good deal to expect from such a strategy, both economically as businessmen and politically as leading citizens; the latter had not too much to lose from schemes of joint ownership in terms of real economic control and something of distinct value to gain if social tensions were thereby reduced and the country's politics stabilised accordingly. In short, an accommodation was possible. The international economic crisis which came to a head in 1973–74 made the task of achieving it more difficult because it forced Manley to increase the pace of change, but again it is doubtful whether it rendered it inherently unattainable. After all, the bauxite levy was supported by the local capitalist class and was eventually accepted with more or less good grace by the American government and the corporations themselves, leaving Manley still balanced on his tightrope.

Two other initiatives, however, tipped him off, both of which could have been avoided. The first mistake concerned the direction of foreign policy, in particular the developing friendship with Cuba and the support, ostentatiously offered, for the latter's involvement in the Angolan civil war. Manley has defended these policies as non-negotiable matters of principle. However, his excessively romantic view of the world underestimated the reality of Jamaica's strategic location in the American sphere of interest. In the last analysis the relationship with Cuba (which in concrete terms produced only minor technical assistance) was not worth the opposition it engendered from the United States. It was not integral to the government's domestic reforms or even the wider pursuit of changes in the international economic order, and in the context of US paranoia about the role of Cuba in the Caribbean would have been better avoided. The second mistake concerned the PNP's readoption of 'democratic socialism' as the ideological framework for reviving populist support for its policies. The choice of label proved to be counterproductive: it alarmed even those local capitalists who had till then supported Manley, further perturbed the government of the United States and appeared to give official approval to the activities of the small left-wing element in Jamaican politics inside and outside the PNP. Other ways of mobilising the Jamaican people existed which need not have incurred such heavy costs.

These two mistakes were critical. Both served to over-politicise what Manley was trying to do and make his policies appear more radical and threatening to established interests than they really were. They aroused unnecessary opposition at home and abroad which not only badly damaged the country's economy but undermined Manley's intended strategy of change. He was unable to rebuild the class alliance on which he had come to power, and although for a time he seemed to toy with the idea of embarking upon a complete disengagement from the capitalist world economy, sustained domestically by the adoption of the politics of class confrontation and internationally by Soviet support, his government was largely devoid of strategy from the moment it became embroiled with the IMF. Manley thus got caught in that no-man's-land between rhetoric and reality in which so many reformist politicians of the Third World have found themselves. His failure teaches the need for consistency of purpose in the pursuit of development. If the strategy is to be fully socialist, it must be followed all the way. This means that the state must be capable of maintaining the productive level of the economy, sufficient outside support must be available, the party must be committed and tightly organised, and a certain level of class mobilisation must be present to sustain the regime through the transition period. Manley never intended to lead Jamaica down this path, knowing how unrealistic an option it was in the circumstances in which he had to work, but some of his language and some of his policies nevertheless suggested that this was his goal, thereby bringing upon his government the wave of opposition one would expect. Equally, if the strategy is to be populist and social democratic in character, as it was in Manley's case, it must be so not only in implementation but in style and presentation. Manley had no need to lose the support of either his local capitalists or the government of the United States, yet he sacrificed both and found that without the former he could not run the economy and without the latter he could not manage the external environment in the way his programme of domestic change required. In the face of the antagonism of these two powerful interests, the degree of manoeuvre left to the Jamaican state was minimal.

In summary, Manley's 'democratic socialist' experiment offers important lessons to proponents of change in the Third World. The task of overcoming, or even adjusting, dependency will never be easily accomplished, least of all in the difficult economic circumstances of the mid- and late 1970s, but success will never be approached unless policy is as consistently implemented and articulated as it is conceived. Manley failed at the end of the day because he got himself into a muddle.

Notes

1 Michael Manley, *Jamaica: Struggle in the Periphery*, London, 1982, p. ix.
2 For a full discussion see Owen Jefferson, *The Post-war Economic Development of Jamaica*, Kingston, 1972.
3 *Ibid.*, p. 285.
4 Steve De Castro, *Tax Holidays for Industry: Why we have to Abolish them and How to do it*, New World Pamphlet No. 8, Kingston, 1973, p. 6.
5 Jefferson, *Economic Development of Jamaica*, p. 285.
6 See Stanley Reid, 'An introductory approach to the concentration of power in the Jamaican corporate economy and notes on its origin' in Carl Stone and Aggrey Brown (eds.), *Essays on Power and Change in Jamaica*, Kingston, 1977, pp. 15–44.
7 Carl Stone, *Class, Race and Political Behaviour in Urban Jamaica*, Kingston, 1973, p. 50.
8 For an elaboration of this model see Carl Stone, *Democracy and Clientelism in Jamaica*, New Brunswick, 1980, especially pp. 91–110.

Jamaica: the 'democratic socialist' experiment

9 *Ibid.*, pp. 94–5.
10 See Terry Lacey, *Violence and Politics in Jamaica 1960–70*, Manchester, 1977.
11 See Carl Stone, *Electoral Behaviour and Public Opinion in Jamaica*, Kingston, 1974.
12 Michael Manley, *The Politics of Change: A Jamaican Testament*, London, 1974, p. 118.
13 *Ibid.*, p. 106.
14 *Ibid.*, p. 215.
15 *Ibid.*, p. 68.
16 Manley, *Jamaica: Struggle in the Periphery*, p. 38.
17 *Ibid.*, p. 221.
18 Manley, *Politics of Change*, p. 130.
19 Manley, *Jamaica: Struggle in the Periphery*, p. 123.
20 Manley, *Politics of Change*, p. 17.
21 *Ibid.*, p. 18.
22 To take a single example, K. Hendrickson, chairman and largest shareholder in the National Continental Conglomerate and Caribbean Communications, was made chairman of the island's new electricity utility, Jamaica Public Service, majority-owned by the state. Reid in Stone and Brown, *Essays*, p. 29.
23 National Planning Agency, *Economic and Social Survey: Jamaica 1979*, Kingston, 1980.
24 Claremont Kirton, 'A preliminary analysis of imperialist penetration and control via the foreign debt: a study of Jamaica' in Stone and Brown, *Essays*, pp. 80–1.
25 National Planning Agency, *Economic and Social Survey: Jamaica 1976*, Kingston, 1977.
26 See 'Bauxite: how the PNP liberals satisfy the local capitalists and betray the masses', *Socialism!* (a Marxist-Leninist journal), 1, 1974, pp. 5–14.
27 See Sherry Keith and Robert Girling, 'Caribbean conflict: Jamaica and the US', *NACLA Report on the Americas*, XII, 1978, p. 21.
28 *Ibid.*, p. 24, table 7.
29 *Ibid.*, pp. 23–4.
30 *Ibid.*, p. 19, table 5.
31 Manley, *Jamaica: Struggle in the Periphery*, pp. 100–1.
32 *Ibid.*, p. 116.
33 *Ibid.*
34 For a full discussion see Anthony Payne, 'From Michael with love: the nature of socialism in Jamaica', *Journal of Commonwealth and Comparative Politics*, XIV, 1976, pp. 82–100.
35 People's National Party, *Democratic Socialism: the Jamaican Model*, Kingston, 1974.
36 See Marlene Cuthbert and Vernone Sparkes, 'Coverage of Jamaica in United States and Canadian press in 1976', *Social and Economic Studies*, XXVII, 1978, pp. 204–20.
37 Made by the former CIA agent, Philip Agee, on a visit to Jamaica in 1976.
38 See Carl Stone, 'The 1976 parliamentary election in Jamaica', *Journal of*

Commonwealth and Comparative Politics, XV, 1977, pp. 250–65.
39 Michael Manley, Speech to the Nation, 5 January 1977, Kingston, mimeo, 1977.
40 Fitzroy Ambursley, 'Jamaica: the demise of "Democratic Socialism" ', *New Left Review*, 128, 1981, p. 82.
41 Manley, *Jamaica: Struggle in the Periphery*, p. 155.
42 George Beckford and Michael Witter, *Small Garden ... Bitter Weed: The Political Economy of Struggle and Change in Jamaica*, Morant Bay, 1980, p. 93.
43 Manley, *Jamaica: Struggle in the Periphery*, p. 155.
44 Stone, *Democracy and Clientelism*, p. 176.
45 According to one account, 'the Soviets were so unenthusiastic about acquiring another ward in the Caribbean that they laid down prerequisites for aid that bore a striking resemblance to the standard IMF regimen for a country in Jamaica's position'. J. Daniel O'Flaherty, 'Finding Jamaica's way', *Foreign Policy*, XXXI, 1978, p. 148.
46 The subsequent discussion of Jamaica's relationship with the IMF relies heavily on the following: Norman Girvan, Richard Bernal and Wesley Hughes, 'The IMF and the Third World: the case of Jamaica, 1974–80', *Development Dialogue*, 11, 1980, pp. 113–55.
47 Manley, *Jamaica: Struggle in the Periphery*, p. 160.
48 A good summary of this argument is provided in Nick Butler, *The IMF: Time for Reform*, Young Fabian pamphlet, London, 1982.
49 Beckford and Witter, *Small Garden*, p. 99.
50 See the interview with Hugh Small, the new Minister of Finance, published in the *Guardian Third World Review*, 1 October 1980.
51 For a fuller discussion see Carl Stone, 'Jamaica's 1980 elections', *Caribbean Review*, X, 1981, pp. 4–7 and 40–3.
52 Girvan *et al.*, *Development Dialogue*, p. 155.
53 For these and other statistics see National Planning Agency, *Economic and Social Survey: Jamaica 1979*, Kingston, 1980.
54 Manley entitles the chapter describing the fall of his government 'Destabilisation triumphs'. Manley, *Jamaica: Struggle in the Periphery*, p. 193.

Paul Sutton

2 Trinidad and Tobago: oil capitalism and the 'presidential power' of Eric Williams

On 28 September 1973 Dr Eric Williams, Prime Minister of Trinidad and Tobago for the previous seventeen years, and founder and political leader of the governing party, the People's National Movement (PNM), throughout that time, announced without warning his intention 'to return to private life and take no further part in political activity'. Seven and a half years later, on 29 March 1981, and as dramatically in the sense of being totally unexpected, Dr Williams died in office of undiagnosed diabetic coma. This chapter concerns itself with these last years and in particular with the consequences of the 'oil boom' which served first as prime cause of Dr Williams's decision to remain in office and then as the foundation of his attempts to reshape the political and economic system of Trinidad and Tobago.

Background: the political system, 1955–73

The necessary background to an understanding of this later period is the years 1955–73, in which Williams's power was first confirmed, then consolidated, and finally corroded. While hard-and-fast dates for this process cannot be projected, it corresponds approximately to the years 1955–61, 1962–66 and 1967–73.

The confirmation of power, 1955–61. As I have argued elsewhere, these are the most important years, 'setting a stamp on all subsequent government and politics in the country'.[1] They are the years of energetic pursuit of government-led modernisation and of charismatic domination of the creole crowd. To realise the former the machinery of government was seen as all-important, which inevitably brought Williams into conflict with the Colonial Office in London and the nascent Federal Government of the West Indies, both of which had claims to a share in power. Williams's character and temperament were utterly against this, and the story of these years is largely of how he confronted and defeated both them and their allies in Trinidad and Tobago.[2] In this the other element, that of charismatic domination, was critical. Williams's ascendancy over the PNM was early confirmed,[3] to be extended into society at large by the adroit appropriation of the flag of the nationalist cause by means of agitation for the return to Trinidadian patrimony of the US naval base at Chaguaramas. Williams's success here rebounded to his advantage when in the general elections of 1961 the PNM increased its share of the vote to 58%, as against 39% in 1956, albeit on a pronounced racial basis.[4] Independence was soon to follow, with a constitution modelled on Westminster and reflecting *de jure* what already existed in practice – the concentration of administrative power in the hands of the Prime Minister.[5]

The principal feature of politics in this period was the ascendancy of Dr Williams. The style was that of 'Doctor Politics' in which 'the Leader is expected to achieve for and on behalf of the population. The community is not expected to contribute much more than crowd support and applause.'[6]

The consolidation of power, 1962–66. These are essentially years of 'state-building'. Williams's main concerns were 'the development and implementation of a foreign policy; the development of a sense of national community; public service reform; and reform of the economy by way of development planning, regulation of labour and capital, and tripartite consultation'.[7] They are marked by the advance of 'technocracy' through the promotion of an administrative elite in the public service, whose prime loyalty lay with Williams and in whose service they were prepared to labour long hours.[8] The centralising tendency of the earlier period thus continued, held marginally in check only by the formal requisites of the Westminster system. Certainly, the PNM was offered no real part in policy-making, and after indifferent attempts at reform in 1963 morale and activity slumped, as Williams himself was to find out in his much publicised 'Meet the Party' tours.[9] Other groups in the country, more vocal in demands, fared no better. In particular there was the failure of an opposition to develop within the framework of the main rival to the PNM, the Democratic Labour Party (DLP).[10] The general election of November 1966 was thus in essence 'no contest', Williams running on the not unimpressive record of government in these years. There was, however, a straw in the wind. Whilst the PNM was again returned in two-thirds of the seats, the percentage of the registered electorate actually voting fell considerably (88% to 66%) as did the absolute number of votes cast.[11]

The corrosion of power, 1967–73. At the end of 1966 Williams could look back on a decade of power with some satisfaction. He had led Trinidad and Tobago to independence; he had set the agenda for social and economic modernisation; and he was without a political rival in a political system which superficially could be described as 'liberal democratic'. If he now consciously thought it was time to take stock, to write the autobiography and the history of the Caribbean long promised,[12] to publicly take a 'back seat' in government, involving himself only as and when absolutely necessary, as in his assumption of the Finance portfolio in 1967, following considerable agitation among business circles in Trinidad, then his reasoning was understandable if ultimately mistaken. The one event which was to shake his confidence, and put in question his administration, was the widespread Black Power disturbances of February–April 1970.

The 'how' and 'why' of 1970 is exceedingly complex. Essentially, however, the explanation would appear to lie in the following:[13]

1. The withdrawal of Williams from the spotlight produced first unease and then a loss of confidence among his supporters, the whole being compounded by rumours of corruption in high governing circles.
2. Concentration of executive power in the Office of the Prime Minister and administrative decisions in the Cabinet resulted in considerable overload at the centre, the result of which was inept and indifferent public administration.
3. Neither Parliament nor established political parties were meaningfully engaged in the political process or showed any likelihood they might be, with the result that political activity was projected on to other organised groups such as religious bodies, trade unions or employers' confederations.
4. The goals of economic reform and social progress were yet to yield tangible results, particularly the reduction of unemployment, which increasingly bore disproportionately upon the young blacks of Port-of-Spain and its environs.

In its programme and actions the Black Power movement was to touch on all these points. It brought people (the crowd) back into politics; identified the agents of political and economic oppression (the Catholic Church, local business and the multinationals); sought the means by which they might be confronted (demonstration and debate); and finally gambled (and lost) on an overthrow of the state. Its miscalculations were Williams's gains, and following the failure of the *coup* in the army he moved swiftly and decisively to restore lost power.

Once again, Williams put himself visibly at the centre of decision-making, taking on the portfolio for almost everything and sacking those Ministers who had been the main targets of criticism. He also sought to restore confidence through, first, the launch of a programme of 'National Reconstruction' which promised, *inter alia*, an immediate reorganisation of the government machinery, to be followed by a fundamental reform of the constitution; second, a revision of the third Five Year Plan to provide more jobs through public works and enlarge the area of national decision-making in the economy; and, third, the involvement of all citizens of Trinidad and Tobago in the difficult task of nation-building. Later in the year these themes were to be expanded and developed as 'The Chaguaramas Declaration',[14] a fundamental statement of policy for the 1970s as 'The People's Charter' had been for the PNM during the 1950s and 1960s. These in turn formed the basis of the PNM's manifesto for the 1971 general election, which it again won easily, but not convincingly, owing to an effective opposition boycott of the polls.[15] Unassailably in power, the question as it unfolded in 1972 and 1973 was whether Williams had either the will or the capacity to stay there.

As it turned out he had neither. In his 'resignation' speech he highlighted three reasons for his departure – Caribbean integration (it was not progressing); the 'state of the nation' (the economy was crumbling and there were guerillas in the hills); and the PNM and its objectives (individualism and undiscipline were rampant).[16] These conclusions but echoed his earlier pessimistic independence anniversary message, which ended significantly with a quotation from Ralph Emerson, 'A nation never fails but by suicide'.[17] The nation was failing from a combination of internal and external pressures to which the political system as fashioned by Williams could not effectively respond. He saw this clearly and, tired of office, decided to go. It was a decision not wholly in character, and thus the cause of much speculation; but it was nevertheless genuinely taken and not contrived.

A summary of the political system to this date would have to note strength and weakness, development and decay, which imparted to political life as a whole an element of predictability but no particular guarantee of durability. That is, while the regime could maintain itself administratively and politically without undue recourse to coercion (though there were elements of this), its legitimacy as expressed in the satisfaction of demands and the mobilisation of support was far from clear. In itself this was not enough to sustain a full-blown 'crisis', though it was sufficient to engender a feeling of unease and uncertainty throughout the nation to which Williams was eventually to respond with his 'resignation'.

Background: the economic system, 1955–73

Parallel to, impinging on and in turn being influenced by, political events were developments in the economy. Attention must be drawn to four areas: the dependence on oil; the decline of agriculture; the limits to manufacturing; and the response of government.

The dependence on oil. The PNM's rise to power coincided with an upturn in the economy. From 1955 to 1961 the economy grew by some 10% per annum in real terms. Thereafter growth fluctuated, being constant between 1962 and 1965; increasing from 1966 to 1968; declining in 1969–70; and recovering in 1971–72.[18] In all this there was one constant – the link with crude oil production. When this was high, the economy was buoyant; when falling, depressed. It could not be otherwise as long as petroleum accounted for such a high proportion of GDP (27%) and of central government revenues (30%).[19] This dependence, of course, was of concern to the government throughout these years. It felt, however, that it could do little to alter the situation, given Trinidad's tiny crude oil production by world standards (0·5% of total production) and its vulnerable position as an intermediate refiner, sustained predominantly by one dominant vertically integrated company, Texaco. Government intervention and regulation were thus at a minimum, confined chiefly to the purchase in July 1969 of the assets of BP Trinidad Ltd for US$22 million, subsequently incorporated in a joint venture as Trinidad-Tesoro, and the passage in December that year of a Petroleum Act designed to consolidate and amend, rather than revise, existing legislation relating to petroleum.

The decline of agriculture. Between 1956 and 1968 the

contribution of agriculture to GDP just about halved (15% to 8%).[20] The chief cause was stagnation in the sugar industry, which dominated the agricultural sector and which operated more or less indifferently throughout this period, its high costs being shielded by sales to protected markets in the UK and USA. External dependence was matched for the government by internal dependence on the industry as an employer of labour. Throughout the 1960s an average of 25,000 persons were directly employed in the industry, the majority of whom were unskilled and middle-aged and therefore without prospects elsewhere. Maintaining the industry was thus an essential aspect of government policy, to which political and ethnic factors further contributed, over 90% of sugar workers being of East Indian descent, organised politically in the DLP. In 1968 and 1970 the government therefore moved to participate directly in the sugar industry when jobs were threatened or it appeared politically expedient to do so, through outright purchase in one case and 51% ownership in another (Caroni Ltd, a subsidiary of Tate & Lyle). Efforts elsewhere to stimulate agriculture, notably the distribution of Crown lands, met with little success.

The limits to manufacturing. Faced with excessive dependence on two commodities, governments in Trinidad and Tobago since 1950 have sought diversification in the development of manufacturing. For most of the period under review the PNM enthusiastically endorsed this, closely following the prescriptions of the so-called 'Puerto Rican model' of development. By the end of the 1960s, however, the returns were meagre, particularly in the generation of employment. From 1955 to 1969 the labour force increased from 267,400 to 368,400 (by 37·8%).[21] 'Pioneer' manufacturing absorbed only a small proportion of this – one estimate being that at the end of 1972 only 18,627 persons were employed in 'assisted' establishments.[22] For this the government had to forgo a loss of revenues estimated by Rampersad at between 1% and 2% of GDP, i.e. 7%–14% of revenues collected.[23] Additionally, the programme directly raised the level of foreign ownership in the economy (local participation in such ventures being minimal), thereby further emphasising dependence, marked not only by remittances of profits abroad but also by the fact that few local raw materials were used.

The role of government. Government action in this period

largely took two forms. One was to set the framework for economic development as a whole through the medium of indicative development planning. The other was to regulate the system in whole or in part through the establishment of a number of pragmatic *ad hoc* institutional devices, the rationale of which was either the fact of independence itself or an immediate reaction to pressing problems.

Four major development plans have been drafted and three have been implemented. The first Five Year Plan (1958–62) concentrated on improving infrastructure to support industrialisation; the second (1964–68) emphasised diversification and community development; the third (1969–73), diversification and the freeing of the economy from dependence on external decision-making. A fourth plan was prepared but not adopted when it became clear that the fortunes of Trinidad and Tobago were being rapidly transformed by the 'oil boom'. It also corresponded to a belief by Williams expressed as early as 1969 that planning was failing:

> In Trinidad and Tobago, results so far do not give much grounds for optimism about the effectiveness of indicative planning. ... Unless satisfactory solutions are found, either attempts at indicative planning will have to be abandoned in favour of more direct measures, or the planning process in many of the developing countries will remain less than fully effective.[24]

The attempt to regulate the system as a whole, other than the budget, focused on financial institutions, labour and capital. In 1964 Acts established a central bank and licensed commercial banks. In 1966 an Insurance Act was passed. Attempts at regulating labour by way of the Industrial Stabilisation Act, 1965, raised a storm of protest,[25] as also did the passage of the Finance Act in 1966, with its alleged 'controls' of capital. The fact that the latter was subsequently modified whereas the former was not was to be a contentious issue in politics in the late 1960s and early 1970s, leading to a degree of alienation of trade union support (notably among the Oilfield Workers' Trade Union, the best organised and strategically most powerful union in the country).

Regulation in part increasingly took the form of direct government participation in the economy. This derived largely from *ad hoc* responses to private-sector retrenchment of labour and more generally from the weak performance of the private sector as a whole. By mid-1972 the government employed more than 100,000 people, making it the biggest and most diversified employer in the country. Some 80,000 of them were employed in the public service and 20,000 others in commercial and industrial activities through government participation in some twenty-one companies with a book-value shareholding of around TT$60 million.[26] In this way the government of Trinidad and Tobago was to obtain the largest stake in the economy of any Caribbean country except Cuba.

All the above, however, failed to alter the essence of the economy. It remained, in 1973 as in 1955, small, dependent and very open. This made it extremely difficult to manage, as Havelock Brewster, among others, has noted. In a quantitative analysis based on 'personal observation and close contact with political leaders and civil servants over the past fifteen years'[27] he demonstrated that over two decades no significant co-variation between the major economic functions in the Trinidad economy – employment, wage rates, exports, import ratio, output, consumption, prices and investment – occurred in the manner prescribed by orthodox economists. A consequence of this is that 'the standard range of institutions – the Plan, the Planning Department, the Central Bank, the Industrial Court', which have been established in Trinidad and may work elsewhere, do not operate effectively in this instance. Given, however, that such institutions 'flourish but do not function', Brewster can only conclude that their rationale is not economic but symbolic – their 'very physical presence is projected as the living symbols that governments govern'.[28] That the hollowness of this should finally have dawned on Williams in the third quarter of 1973 when foreign exchange reserves were the equivalent of less than two weeks of the country's imports is, of course, far from coincidental.

'Resignation' and 'renewal': 1973

In 1973 the new offshore oil fields developed by the US transnational Amoco came substantially on stream, offsetting the serious decline in crude oil production of the previous few years and returning the level of production to that recorded in 1968.[29] In January 1973 Amoco paid approximately US$0·50 tax and royalty per barrel of oil; in December 1973 it was paying US$4·69, and as new prices and taxes posted by Trinidad and Tobago took effect from 1 January 1974 this was practically to double again. With this the central government revenues of Trinidad were transformed. In 1973 the total was TT$591·2 million, the following year TT$1397·7 million, and by 1978 it was TT$3,226 million.[30] 'Money is no problem' was how Williams was to explain the new situation in 1976. More than other factors it was the desire to exploit it fully that prompted him to remain in office at the end of 1973, thereby ushering in a new phase in the political and economic history of Trinidad.

This, of course, is not to discount political factors as operative in his decision to remain. They undoubtedly played a part – particularly in respect of the confusion and dissension caused within the PNM by his 'resignation'.[31] However, there is no reason to believe that politics alone would have prompted a change of heart, though Williams was later to capitalise on the situation which his 'resignation' had caused. Indeed, it is precisely such 'opportunism' that is most characteristic of him in respect of his actions then and earlier. If it could be firmly grounded on an economic base, which from his earliest days he had recognised as 'ultimately determinant' of political action,[32] then so much the better. The 'oil crisis' of late 1973 promised precisely this, and it was towards defining policy in this area that he first turned.

Oil capitalism, 1974–76

The changed fortunes of the nation in 1974 as compared to 1973 are evident in many government pronouncements but nowhere do they have greater contrast than in the independence messages of these years. Whereas 1973, as we have seen, ended with an ominous warning, the conclusion in 1974 was positively euphoric: 'let us say, with pride but yet with humility, we are going well, and may God bless our Nation'.[33] More than this, however, the speech was noteworthy for spelling out how the oil 'windfall' was to be used for the development of the country:

> In our case oil means (a) a large number of additional permanent jobs through downstream petroleum operations or new industries based on petroleum, (b) greater national ownership of our national resources meaning specifically greater national decision-making and the local utilization and diversification of products which we formerly exported, (c) larger allocations for our domestic services, (d) more rapid progress towards Caribbean integration, to supplement our own domestic efforts at greater self sufficiency.[34]

In the furtherance of the second of these goals the speech also marked the purchase of the holdings of Shell Trinidad for TT$93·6 million, a price regarded as inflated by some but of incalculable symbolic significance for the government.

Taking in turn the four headings spelled out by Williams, the following observations on the economy in the period 1974–76 can be made.

Energy-based industrialisation. The government's response to the 'oil crisis' was the establishment of an Energy Secretariat to coordinate all matters relating to petroleum, and the despatch overseas of three separate high-powered diplomatic missions to examine the ramifications of the 'oil crisis' as it was developing and was likely to affect Trinidad and Tobago. Simultaneously Williams, in a series of 'addresses to the nation' in the first half of 1974, was to report that Trinidad and Tobago was being courted by a number of large transnational corporations intent on using the country's abundant natural gas reserves to establish a manufacturing presence on a joint-venture basis with the government. In October 1974 he was to list this activity as follows: agreement with Texaco in respect of two petrochemical plants (51% government holding); work in progress on a second ammonia plant as a joint venture with W. R. Grace (51% government holding); discussion of an aluminium smelter to be located in Trinidad, based on alumina from Jamaica and Guyana (34% government holding); discussion on a sponge iron plant to be located in Trinidad, based on iron ore from Brazil; and examination of proposals from Amoco, Hoechst, Beker, Kaiser, Mississippi Corporation and Tenneco for additional fertiliser plants (all 51% government holdings). Additionally, the government was to embark on a major electricity expansion programme and to lay a natural gas pipeline to a new purpose-built industrial estate at Point Lisas.[35] While not all these ventures were eventually to get off the ground, the majority were. Evident within them is not only an expansion of the Trinidad state but also the directing hand of Dr Williams, eventually made explicit with his resumption of the Finance portfolio in 1975.

National ownership. It is important to remember that the goal of greater national ownership came before the 'oil crisis' and was the object of a comprehensive White Paper in 1972. It was to be realised alongside new foreign investment, to which it was seen as complementary rather than substitutive, i.e. the mixed economy was to remain as before: all that had changed was that the government was now expected to play the role of a prime mover in its development. Nationalist rather than socialist objectives were the motivating forces, for which more than any other figure Williams was responsible. Accordingly, pragmatic considerations decided the national interest in each case. This is well illustrated by the example of Texaco. In May 1974 Williams indicated that the government was interested in acquiring a stake in their activities in Trinidad. Thereafter he proceeded to make haste slowly. 'Serious' negotiations did not begin until nearly a year later, by which time the indications were that he was changing his mind. In two speeches in April 1975 the whole question of national ownership was raised rhetorically and the answer given that it was not always expedient to proceed, especially at that moment in the oil industry.[36] Thereafter Texaco slipped from view. This did not mean, however, that the government was not proceeding on other fronts. In 1975 a second White Paper on public-sector participation in industry was published which noted that government holdings had increased to approximately TT$209 million, covering a wide range of activities. The most notable additions in that year were the purchase of the remaining shares in Caroni to make it a wholly government-owned company.

Domestic relief. The 'oil crisis', of course, coincided with the onset of recession and double-digit inflation in the Western economies. To this Trinidad and Tobago, with its very open economy, was particularly vulnerable, and the government quickly moved to cushion the effects through the deployment of various devices to raise standards of living, which elsewhere in the Commonwealth Caribbean were being seriously eroded. Its action took three principal forms. The first, affecting everyone, was the implementation of a comprehensive system of subsidies. From 1974 to 1976 they amounted to some TT$492 million, allocated under the following heads: basic food (17·8%), agriculture and fishery (1·8%), welfare (13·0%), utilities (37·2%), petroleum products (29·6%), other (0·6%).[37] The second was a series of measures to reduce taxes, hence primarily affecting the employed. The areas covered were income tax (relief and rebates); motor vehicles and motoring; purchase tax (refrigerators, freezers, stoves and garments); and stamp duty (land transfers). The amounts of revenue estimated as forgone were, in 1974, TT$36·8 million; 1975, TT$41·6 million; 1976, TT$26·4 million.[38] Finally, action was taken to expand the various programmes designed specifically to relieve unemployment. The problem here, as Williams frankly recognised in 1975, was that jobs were traded for votes, with the PNM and its representatives, not surprisingly, being the principal beneficiaries. This he condemned.[39] However, little was subsequently done to remedy the situation, and the question must be asked how different in intent this was to the other measures outlined above, especially given the well known consumerist ethos of West Indian society.

Caribbean integration. Williams's 'addresses' on the 'oil crisis' in the first half of 1974 were peppered with references to how oil could be the fuel for advancing Caribbean integration. As it turned out his hopes were not to be realised, and the forward momentum of 1972, already slowing in 1973, came to a full stop by 1977, the year which 'witnessed the near total collapse of the Caribbean Community Treaty'.[40] All that need be said is that, whatever transpired afterwards, Trinidad and Tobago was not to blame up to that date.

The combined effect of the above measures, of course, has been considerably to widen the role and scope of the state. By 1976 the public sector, directly and indirectly, was employing over half the labour force and there was scarce any area of economic activity in which its remit did not run. Williams was not unaware of the problems this posed, but naturally enough solutions were not easy to find. In the event, two appear to have been adopted. One – less a solution than a postponement – has been the medium of Special Funds. Every budget since 1974 has set aside considerable revenues in a number of such funds, each earmarked for a specific purpose and designed to effect long-term development. At the end of 1976 twenty-three such funds had been established and appropriations totalling TT$1,420 million had been made.[41] The other has been the promise of future divestment of government holdings. This is in accord with a philosophy which sees state ownership largely as a temporary form of trusteeship pending the emergence within the economy of the 'small man'. Since the latter had still to be realised, caution was to be exercised – especially given the rapacious nature of the business class in Trinidad and Tobago.[42] Here, starkly laid out, is Williams's personal preference and strategy for development. Simply put, it sees the future economic development of Trinidad and Tobago as lying in the hands of a national middle class at present in formation but not yet independently formed. The task of the PNM, through control of the state, was to see that it did.

Political reconstruction, 1974–76

The system of specifically political power that Williams had built to 1973 rested on three foundations: the Westminster system, the PNM and the public service. In all three Williams dominated by virtue of office and example, and in all three the habit of concentrating the power of decision in his own hands was very evident. As he himself said in the 1971 general election campaign, 'I'm the one who has power here. When I say "come" you "cometh", and when I say "go" you "goeth".'[43] What, then, is most extraordinary about the period 1974–76 is that Williams believed he did not have enough power, or perhaps more exactly, given the experiences of 1970–73, that his power was insecurely based. He therefore systematically set about breaking all three 'foundations', reconstructing them in a way designed to give him absolute personal power by the time the general elections were fought in September 1976. The way in which he did so is examined below.

Constitution reform. On 18 June 1971 the establishment of a Constitution Commission was announced in the Speech from the Throne, and in January 1974 the Commission duly presented its report and draft constitution. Its main recommendation was that 'We are of the view that the Westminster model in its purest form as set out in our present Constitution is not suitable to the Trinidad and Tobago society'.[44] Consequently, in its draft constitution, it proposed a number of changes, some of which were significant departures from the existing constitution, though none was cumulatively enough to question in any fundamental way the Westminster system as formally established.[45] Thereafter the issue was put to the public in a largely inconsequential debate until Williams pronounced on it in a massive seven-hour speech to the House of Representatives in December.

A speech of such length obviously covered much ground. However, in the light of later developments one aspect appears fundamental – his accusation that the Constitution Commission had 'an illogical obsession proceeded with logically to reduce the powers of Parliament, to create confusion and to break up the centralization of Parliament, Cabinet and the political party',[46] i.e. a belief by Williams that the Commission's motivation was partisan and that by design it sought no less than the destruction of the system he had created, to the immediate disadvantage of himself and the PNM. He therefore refuted almost in their entirety its investigations and recommendations, paying particular attention to the question of proportional representation, the adoption of which he utterly opposed, either in whole or in part, on the grounds that it would 'dissolve the present PNM majorities'.[47] Not stated, though obviously derived from it, was the concomitant proposition that it could also lead to an end to Williams's sole leadership in Parliament (i.e.

coalition government) or, even worse, Opposition. Needless to say, and on his past record, this was something Williams would not easily have been able to accommodate. He therefore proposed that note only be taken of the reports of the Commission and that Parliament itself proceed to draft a constitution more acceptable to the country as a whole.

This exercise took up the whole of 1975, and finally, on 12 March 1976, Williams moved the adoption of the new constitution in the House of Representatives. It contained a number of novel features. Trinidad and Tobago was to become a republic with a President as head of state; the Senate was to remain, but whereas the previous constitution had limited the numbers of Ministers that could be drawn from the Senate, no such limitation was now to be imposed; eighteen was to be the age of majority, and voting was to be by ballot box and not by voting machine; an Ombudsman was to be appointed; and an Integrity Commission to be provided for, its exact powers to be determined in future legislation.[48] Much of this was uncontroversial, the measure which raised the greatest comment being that related to the Senate. The Constitution Commission had recommended the abolition of the Senate. Williams rejected this out of hand, favouring instead an enhanced role for the Senate in Parliament and in government. The stated reason for this in so far as one is discernible was to cast the net for talent as wide as possible. At the same time, however, it must be noted that a consequential effect is a substantial increase in Prime Ministerial power. That is, while members of the House of Representatives have a measure of independence from the government by virtue of their seat, senators are removable at will. In such a situation pressures to conform, it might be adduced, are more likely to prevail than vociferous dissent.

The public service. In his address to the sixteenth Annual Convention of the PNM, Williams hinted that all was not well in the public service. It was still imbued with 'the techniques and procedures of colonialism, as well as the mental outlook associated with it ... [and] with the exception of a very few public servants is just not geared to the execution of the responsibilities which the scope and pace of present development are imposing on it'. He went on:

> If all the advice tendered to us recently had been accepted, we would have found ourselves leaning over backwards to support some group's political line; we would have crawled on our belly to achieve a subordinate role in one particular organization; we would have broken

off diplomatic relations with this country or that; we would have gone out of our way to show our support for this leader or that; and all of you would have awakened one morning to find that overnight you had become the Protectorate of some country outside on the basis of demands stated with brutal precision.[49]

A year later this was expanded to form the substance of a considerable attack on what he termed 'a small ambitious minority of senior civil servants',[50] among whom were included D. Alleyne, then Head of the Civil Service and Permanent Secretary to the Prime Minister.

The choice of 'target' here was far from fortuitous. If Alleyne was not safe from public censure and disgrace, then who was? Equally, the thrust of Williams's attacks was well chosen. It was on the activities of the Energy Secretariat in the early part of 1974. This body, he charged, had illegally by-passed the Tenders Board; transferred individuals from substantive positions in various Ministries without approval; offered partisan and unsound advice in the interests of foreign companies and governments; and, finally, sought to involve the Prime Minister himself in its nefarious activities.[51] In other words, a government within a government was in the making, and Trinidad and Tobago 'stood close to take-over by a technocracy, only wanting someone to convey an aura of respectability by chairing its committee'.[52]

If so, then Williams's own part in this cannot be ignored. He early courted technocracy, and his initial decision to establish an Energy Secretariat was in no way surprising or in any sense an unusual departure from existing practice. What was new, of course, was the context. The 'oil boom' necessarily placed a premium on technical advice and the 'independence' that went with it. Hitherto, Williams had been able to manage the latter either by direct oversight (location of such advice principally within the Office of the Prime Minister); or by political control at one remove (the device of the Minister of State); or simply by adroit personnel management (which included 'banishment' from the Prime Minister's favour). Now the very scale of operations envisaged threatened this. Temporarily, Williams was disarmed. Once the future pattern of the development of hydrocarbon resources had been decided, however, he was able to act decisively. One means, disciplinary vigilance, signalled that, if his reliance on technocracy had increased, it was not to be unconditional. The other, later developed extensively, was the institution of multiple portfolio-holding by subordinate Ministers. This led to a situation which, if not exactly analogous to the politics of a Renaissance court, was not far removed from it either. With everyone watching everyone else and everyone entitled to make a decision even if they chose not to do so, administrative efficiency was clearly discouraged at the same time as central political control was effectively exerted.

Party control. The PNM had been built and developed under the unquestioned leadership of Williams. His association with the party was its guarantee of success, and little was done to develop it either organisationally or ideologically. When, therefore, he 'resigned' in 1973 'confusion glorified' reigned, only finally to be dispelled by his 'return'. This, initially, was not explicitly on any understanding of change in party procedure or organisation. It emerged only later, and specifically in the context of the 1976 general election.

In a speech to an 'elections preparations' meeting of the PNM's General Council on 23 May 1976, and to the astonishment of those present, Williams announced:

> The electorate wish no part whatsoever of the majority of the PNM incumbents and nominees, no matter who else is satisfied with them. ... Let it be understood now, once and for all, by all and sundry – the Political Leader has not the slightest intention of encumbering himself, yet again, with these traditional party millstones, unable to speak properly, knowing nothing of basic issues facing country and world; incompetent for higher responsibilities which ultimately fall on the Political Leader's shoulders, unable – unbelievable though it sounds – even to seek to assist their constituents in difficulty who further turn to the Political Leader and interfere with his attention to his formal, public, national responsibilities. ...
>
> Either the nominees remain and you get a more appropriate Political Leader, or the Political Leader remains and you get more appropriate candidates for whom no apology need be made. There is no other alternative.[53]

As in 1973 the initial reaction within the PNM was confusion and dissension. But Williams was eventually to have his way, and fifteen new nominees were presented (along with five of the 'millstones' for whom he subsequently refused to campaign). More to the point, all had accepted the new conditions for candidates which Williams had long advocated[54] but previously been unable to impose on the party. These required, *inter alia*, the declaration of assets and liabilities; a cast iron commitment to pay a percentage of salary to the PNM; and, most important, the provision of an undated signed letter of resignation to the Speaker of the House of Representatives, to be held personally by

Williams. This last condition, of course, was the disciplinary measure *par excellence*. Williams's own justification of it was couched in such terms – past practice had shown it to be absolutely necessary in such an essentially undisciplined society as Trinidad and Tobago.[55] His main opponent within the PNM, Karl Hudson-Phillips, saw it differently, however. He declared he would not sign such a letter (and hence not contest the election), declaring that it 'could lead to the subjugation of the will of Parliament to a single individual and to a frustration and indeed subjugation of the will of the electorate'... It was 'a sure recipe for dictatorship'.[56] On this question Williams wisely kept his counsel; after all, he had achieved what he intended. Not only was he leader of the PNM but when occasion demanded he could also be above it.

In the late 1960s, as we have seen, Lloyd Best coined the phrase 'Doctor Politics' to describe the political system in Trinidad and Tobago. His description of the evolving situation as it emerged in 1976 is equally apt:

> We have returned to Crown Colony Government with a vengeance. A Governor (masquerading as a Prime Minister) as the only responsible official; a Legislature as decoration pure and simple; an Executive Council of hand-picked officials, some from the Lower, some from the Upper House but all at the mercy of the Chief Executive.
>
> It is the Trinidad and Tobago variant of Latin American Presidential power....[57]

All that was lacking was a new mandate from the people. This Williams now sought and won. He did so in an extraordinary campaign which saw him launching attacks on all around him, friend and foe alike.[58] He did so because he knew, whatever else had transpired, that he retained the faith of a significant percentage of the electorate. This was mapped at the time in a national opinion survey. Asked, 'Do you on the whole approve of Dr Williams's performance as Prime Minister?' 62% of respondents said 'yes', 22% 'no', with 16% having no opinion. More pointedly, asked 'Who would you like to see as the next Prime Minister?' 35% of respondents expressed a preference for Williams – a level of support greater than all his opponents' combined.[59] The PNM, not surprisingly, won the election easily, being returned in twenty-four of the thirty-six seats contested, with the support of 54% of those who voted. Again the turn-out was low, only some 56% of registered electors voting.

The final administration, 1976–81

In his final administration Williams sought to harness the political gains of the previous three years to the task of rapid economic development. As in 1962–66 this was an attempt at consolidation of power through the medium of 'state-building', but whereas then the task had been essentially politico-administrative in character, now it was overwhelmingly economic, shown most forcefully in Williams's retention of the Finance portfolio throughout this time. This permitted him, by virtue of office, not only to give general direction to economic transformation, but also to establish the priorities, notably the creation of a petroleum-based and energy-intensive industrial sector and the use of Special Funds as instruments of development. To carry this out a considerable concentration of political and administrative 'talent' was gathered together. At all times, though, Williams was careful to retain formal authority, further manipulating the political system to this end and appealing direct to the people as 'father' of the nation when occasion arose. A presidential style thus evolved against which opposition often appeared futile. However, it was not entirely absent, and, much as power was being consolidated, it was also being eroded, especially in the very last years of his administration. Indeed, at the time of his death these contrary tendencies were finely balanced, imparting to the political process as a whole that degree of uncertainty which had characterised Trinidad and Tobago in 1970 and 1973.

Industrialisation. Williams's first commitment was to industrialisation. The basis for this was natural gas, which Trinidad and Tobago possessed in abundance, being 'one of no more than ten countries in the entire world to have this precious commodity surplus to needs'.[60] The export of gas via the construction of a liquefied natural gas facility was thus a distinct possibility, and feasibility studies and investment arrangements to this end continued throughout the period. However, of far greater importance was the utilisation of gas locally to link the petroleum sector directly with industry, and during Williams's administration three major projects were to be realised in this area – a fertiliser joint venture with W. R. Grace (TRINGEN), a fertiliser joint venture with Amoco (FERTRIN) and the establishment of a wholly government-owned iron and steel plant (ISCOTT). The estimated cost of all three, at the end of 1979, was some TT$1,700 million,[61] to which should be added substantial infrastructure costs relating to the development of the Point Lisas industrial estate of not less than TT$200 million directly and millions more indirectly.

Such massive investments were unprecedented in Trinidad, and it was only to be expected that difficulties would arise. To meet them a Co-ordinating Task Force under the chairmanship of a long-standing associate of Dr Williams, Professor K. Julien, was appointed in 1975. This agency soon assumed the central direction of the industrialisation effort and, despite setback and delay, retained Williams's confidence throughout, being reorganised as the National Energy Corporation in 1979. Williams's role was thus not one of day-to-day direction (this falling to Mahabir as Minister in the Ministry of Finance and Minister of Energy and Energy Based Industries) but of 'propagandist at large' for the entire effort and final arbiter of difficult decisions. Among the most controversial of these were the introduction of government-to-government arrangements in the 1979 budget speech. The background to this was weak performance by local contractors in fulfilling the tenders awarded to them. The solution proposed, in a novel twist on the 'industrialisation by invitation' theme, was for the government to approach selected foreign governments and ask them to sponsor a particular project by (*a*) designating competent national firms (public and private) with which the government of Trinidad and Tobago would enter into agreements; and (*b*) guaranteeing or assuming responsibility in some explicit and practical way for the satisfactory completion of projects undertaken by the designated firms. Eventually some forty-one projects were identified under these arrangements, and inevitably those involved were drawn overwhelmingly from transnational corporate interests in North America and Western Europe. The paradox of energy-based industrialisation in Trinidad and Tobago was thus revealed as acute dependence in the short run in the hope of independence in the long run, a cost amply illustrated by Williams at the end of 1978 when he noted:

> United States firms and institutions dominate our energy-based industries. We have joint ventures with United States firms for fertilizer and liquefied natural gas, and are negotiating in respect of methanol, aluminium, petrochemicals. The United States Export-Import Bank has loaned us over US$200 million in respect of FERTRIN, ISCOTT and the new Point Lisas power plant. US commercial banks are heavily involved in the financing of FERTRIN, TRINGEN, ISCOTT and the power plant. US firms have won in the international bidding for such contracts as the power plant and major water projects.[62]

Needless to say, such developments have met with considerable criticism inside Trinidad, where it has been argued that for all the costs involved the returns may well be slight, with risks outweighing certainties, particularly in securing adequate foreign markets for which production was necessarily geared. It was also noted that only limited employment opportunities would arise from such industries and that 51% control in joint ventures with transnational corporations was in essence no control at all. Against this Williams simply counterposed the alternatives – leave the gas in the ground, flare it, or, more realistically, export it, in which case he gave a warning:

> some ten years ago the Government, with the knowledge that extensive hydrocarbon resources were discovered off the East coast, was presented with a deal signed and sealed in foreign boardrooms, a deal which effectively said to us: Company X was going to sell Company Y our Trinidad gas which Company X had discovered. It was going to sell it at 25 cents per thousand cubic feet, liquefy it, and ship it to a country where the President had committed himself to provide cheap gas for the population. A deal signed and sealed, ready for execution. We almost did execute it in the moment of our exhilaration: we had gas and there were people interested in buying it from us. What account of our stewardship would we have provided today if following the advice and recommendations of many, we had proceeded along that path – 25 cents for gas which can now sell across the Mexican border at US$4.00 per thousand cubic feet. Millions of cubic feet of gas leaving our shore every day while we have deficiencies in the gas supplies for our electricity. No iron and steel, no more fertilizers, no methanol, no urea, no possibility of aluminium.[63]

In other words, the arguments for energy-based industrialisation were irrefutable and the programme would proceed come what may.

Special Funds. In 1974, as noted earlier, the government set aside quinquennial development planning for a system of annual review. In practice this has had two elements: expenditure incurred in what continued to be referred to as a Development Programme under which head some TT$1,818·9 million was spent to the end of 1980; and appropriations to Special Funds for long-term development amounting to some TT$7,503·3 million in the same period.[64] This latter figure, in both absolute and relative terms (37% of government revenues 1974–80), was by any account a considerable commitment to future growth in Trinidad and Tobago. What form it will take may in part be gauged from a sectoral breakdown of fund allocation and expenditure to date.

Of the total appropriations, 26·2% were made to the 'economic' sectors, 22·4% to the 'social' sectors, 43·4% to 'infrastructure' and 8% to 'other' sectors.[65] In the economic group, agriculture's share was 2·7%, compared to petroleum's 16%, manufacturing's 0·7% and that for commerce and finance of 7%. On this basis 'petroleum', in its widest sense, will not only carry within it significant government participation but will also continue to be a prime motor for growth. In the 'social' category, education received 7·4%, health 1·7% and housing 10·4%. The high figure for housing here represents the government's response to spiralling costs which had put home ownership beyond the reach of even the moderately well-to-do. With respect to the appropriations for infrastructure, approximately half are intended for transport (air, land and sea), while the provisions for water and sewerage amount to 9·7% of total appropriations. The intention here was clearly to make available in Trinidad and Tobago all the necessary trappings of a modern, namely Western, society at its most advanced. Finally, in the category 'Other Funds', two funds relating to building projects and land acquisition made for nearly 40% of total appropriations under this head.

Expenditure from the funds amounted to TT$3,482·3 million to the end of 1979.[66] It was allocated as follows: 'economic' sector, 35·2%, the two highest-spending funds being the Petroleum Development Fund (25·6%) and the Participation in Commercial Enterprises Fund (6·5%); 'social' sectors, 16·3%, the two highest-spending funds being the Education Fund (6·7%) and the Housing Fund (6·5%); infrastructure, 45·7%, the principal expenditures being the Water Resources Fund (12·2%), the Air Transport Fund (7·3%), Roads Fund (7·2%) and Point Lisas Infrastructure Development Fund (6·5%); and 'Other Funds', 2·8%. Patterns of expenditure, as might be expected, are thus similar to patterns of appropriation, and the combined effect of both has been to increase considerably the share of GDP accruing to the construction and financial sectors. Alongside this, of course, has been the deepening and widening of the government's role in the economy, to the extent that one observer has suggested that 'government's current and capital expenditure taken together may now be well in excess of 50% of Gross Domestic Expenditure as compared to 15% in 1959'.[67] The very real power this must give the Minister of Finance needs no emphasis. Williams, of course, was well aware of this and he used it as the principal foundation upon which he further elaborated the presidential style.

Presidential style. In the flush of victory after the 1976 general election Williams told the PNM, 'As your Political Leader, I remain at your service for as long as I may be needed or available. I have helped in twenty years to make PNM great. I shall do all I can to assist PNM in continuing to prevail.'[68] Three years later he was to reveal to the party how he was fulfilling this pledge:

> I have in effect taken the back seat – taking care particularly not even to appear to compete with our new head of state, the President of our Republic ... and taking equal care to allow my Cabinet colleagues to occupy the centre of the stage so that the public can judge between them. ... I have taken care also to avoid even the semblance of foreign interference or local lobbying ... avoiding like the plague all that wining and dining which has traditionally been associated with West Indian society and politics. The Government, the Cabinet, speaks on this that or the other as its duty to the country, the Prime Minister's utterances are reduced to the irreducible minimum. ... So I reserve my statement and activities for general issues and fundamental national objectives and for re-affirming the basic philosophy of the PNM.[69]

Withdrawal, not engagement, was the essence of the Williams presidential style, leading him to live the life of a recluse in the last few years before his death. Administration and government were carried out by others – authority, however, remained vested in Williams' hands, and he took care to ensure that it should.

First and foremost, this meant control of the party, and the principal strategem here was a steadfast refusal to designate a successor. This, of course, capitalised on the experience of 1973 and proved an effective whip, keeping all but the most recalcitrant in line. Leadership issues thus did not publicly emerge in this period and the party made do with the 'temporary' arrangement of 1970 – the provision of three Deputy Political Leaders, namely Mahabir, Chambers and K. Mohammed. At the next level, the legislature, discipline was effected through the medium of the undated letter of resignation coupled with the passage of the Constitutional (Amendment) Act of 1978. This provided that 'a member of the House of Representatives shall vacate his seat in the House where, having been elected with the support of a party, he resigns from or is expelled by that party'.[70] The introduction of this legislation followed immediately upon the defection from the Cabinet and the PNM of the former Minister of Works, Hector McClean, and was clearly intended to give 'teeth' to the 'letter of resignation' held by the Prime Minister. Its effect overall is difficult to judge, since it did not deter two 'millstones' – Carlton Gomes and Brensley Barrow – from writing to Williams to ask for the return of their letters; and the legislation itself was later declared unconstitutional by the Appeal Court of Trinidad and Tobago. At the same time, however, it did give force to Williams's assertion that 'When a constituency chooses its member of Parliament, he is not just a Member of Parliament, he is a PNM Member of Parliament',[71] and the record shows no others following McClean's example. If only negatively, then, it might be concluded that the measure acted as some form of constraint. Finally, at the level of the individual party member, disciplinary procedures could be invoked. This was the course followed most emphatically in the cases of Karl Hudson-Phillips and Ferdi Ferreira, both senior members of the PNM and both suspended in March 1980 for publicly criticising Williams's address to the 1979 Annual Convention of the PNM as 'totally irrelevant'.[72]

Beyond the PNM, Williams could, of course, appeal to the country direct. In these last years he chose to do so largely through the medium of live broadcasts, favouring especially the set pieces of the annual address to the PNM convention and the annual budget speech. These typically became occasions for wide-ranging reviews over the whole spectrum of government and the economy. In them Williams not only revealed his erudition but underlined and emphasised his unrivalled claim to authority as visionary and technical expert rolled into one. An image of the statesman in action – remote, omnipresent, but above all benevolent – was thus assiduously fostered, the benevolence stemming directly from his position as trustee for the nation in his dual capacity as Corporation Sole[73] and Minister of Finance. The rhetoric employed here was an identification with the interests of 'the small man', whom Williams claimed constantly to represent and defend. At one level this meant promotion of the policy of divestment of government holdings, with its promise of shareholding for all, either directly or indirectly. At another it implied continuation of the policy of subsidies and tax concessions. The extent of these to 1976 has already been indicated. Thereafter they continued to escalate, with subsidies reaching a staggering level of TT$2,420 million for the period 1977–80 inclusive. Allocations under this head were as follows: basic food (14·7%), agriculture and fisheries (2·6%), welfare (10·7%), utilities (39·6%), petroleum products (27·3%), and others (3·8%).[74] Income tax concessions were also granted in each budget, with the amount of revenue forgone estimated as TT$24·8 million in 1978; TT$16 million in 1979; TT$13·1 million in 1980 and TT$32 million in 1981.[75] From

1978 onwards the wisdom of such measures came to be questioned, both inside and outside Trinidad and Tobago; the government responding with the appointment of a Committee to Review Government Expenditure which reported in October 1978. Its major recommendations were that no new welfare programmes be implemented and that there should be a moratorium on further tax reductions. While in subsequent budget pronouncements Williams was to pay tribute to the work of the committee, there is little evidence of any real attempt to implement its recommendations or grapple seriously with the problem posed. 'Giveaways' thus continued to characterise government fiscal policy, and it was finally left to Williams's successor to call a halt with his sober presentation of the 1982 budget, which ended ominously with the words, 'the fête is over and the country must go back to work'.[76]

The erosion of power. In the 1976 general election campaign Williams invited those who did not like the PNM's policy to 'emigrate to the Bronx and clean toilets'.[77] This contemporary update of an early 1960s remark that those who did not like what he was doing 'can get the hell out of here' was to find an immediate test case in Tobago, where the electorate had just rejected the two PNM incumbents and returned to the House of Representatives Williams's former deputy, A. N. R. Robinson. His vigorous presentation of the case for internal self-government for Tobago constituted a vexatious problem for the government over the next five years. The question of whether internal self-government should be conceded was not itself an issue, the government agreeing to it early in 1977.[78] The form it should take, however, was, and led to inordinate delay and tortuous proceedings, so that the Bill granting internal self-government did not reach the House of Representatives for debate until 12 September 1980. As enacted it provided for a fifteen-member Tobago House of Assembly (twelve elected and three appointed) charged with implementing in Tobago government policy in respect of finance, economic planning and development, and the provision of various local services. These were powers considerably greater than any local government in Trinidad, yet at the same time they were less than the self-government originally envisaged, and the legislation carried a sting in its tail. This was the provision, ostensibly underlining the unique powers of the Assembly, that members of the Senate and House of Representatives were barred from belonging to it. For Robinson, for whom no doubt it was intended, it left a particularly cruel choice of alternatives which he was faced with almost immediately when elections were held for the Assembly on 24 November. Robinson chose to contest, resigning his seat in the House of Representatives to do so. His Democratic Action Congress (DAC) was returned in eight of the twelve seats in what was regarded as a high poll (63%). Williams took no part in the campaigning and, all things considered, the PNM retained a substantial foothold, winning 45% of the votes as against the DAC's 53%.[79] This served to underline that while Tobago had been an irritant and running sore for the government, it had never led to a haemorrhage of support. What was questionable, however, was whether the same could be said of other issues, of which corruption and accountability were to receive prominent coverage.

The belief that 'corruption is rife among those who hold high political office'[80] has been a staple of Trinidadian politics before and since the PNM came to power. It has fed largely on rumour and gossip, with the added spice of the odd uncovered instance of administrative impropriety, usually at lower levels.[81] Williams and other high officials in the PNM and public service have not escaped accusation, though proof has not been forthcoming. In 1980, however, a number of scandals surfaced and it appeared that at last evidence might be found to implicate, if not Williams himself, then a number of those around him. Easily the most embarrassing were the so-called 'DC9 scandal' and the 'race track scandal'. In each case it was later shown that 'questionable payments' had been made to former high officials of the Trinidad government by US-based transnational companies to obtain lucrative government contracts, and in both instances the matter was thoroughly aired in the Trinidad and Tobago press, to the delight of the opposition. Williams's reaction was to attempt to distance himself from any controversy, seeking to avoid making any specific comments and reserving his interventions to general statements on the themes of accountability and integrity. On this front, however, far more was promised than was ever achieved, raising doubts as to his desire to prosecute the matter vigorously. Two examples may be cited. One concerns the Integrity Commission, provided for in the 1976 constitution. By 1981 it had still not appeared, and in that year all he was promising party and nation was that more would be said 'in due course'.[82] The other was to show that if Trinidad and Tobago was not entirely above reproach on this matter, then nor were others. In 1979 he thus warned:

> We must be on our guard against the agents and promoters, international and local, who seek to peddle their services and create by their very presence an environment which is a congenial breeding ground for corruption. We must guard against those who would seek to have us smeared and set about in a very deliberate and subtle manner to achieve that end. We must guard against the impression that the mere fact we are a young and developing country automatically qualifies us for a handout over or under the table. I am aware of this growing breed of agents and promoters. What are the services they offer and to whom? I may not know all the answers but I have knowledge of some.[83]

In other words, corruption could not be condoned, though it might be understandable in the circumstances. A constant temptation to many, it was 'a sword of Damocles' hanging over their heads and giving Williams an advantage in the short term, with effects clearly corrosive to the body politic in the long run. That the reckoning itself might be sooner rather than later was indicated sharply with the emergence of a new party on the political scene.

On 19 April 1980 K. Hudson-Phillips, the designated successor to Williams at the time of the 1973 resignation crisis,[84] launched the Organisation for National Reconstruction (ONR), in Woodford Square, Trinidad. The timing was significant. It was but two days before the local government elections in Trinidad, which were widely expected to confirm, and indeed apparently did, the collapse of the United Labour Front (ULF) as a credible opposition within the country.[85] It also anticipated, by a matter of weeks, Hudson-Phillip's expulsion from the PNM for criticisms of Williams, the party and the government. The ONR was thus conceived simultaneously as a response to a political vacuum and as a vehicle for the ambition of its leader. Over the coming months it was to attract widespread publicity and generate significant support, so creating difficulties within the PNM camp. Of these the most important was the acknowledged fact that the ONR was in many respects a 'clone' of the PNM, advocating the same policies and seeking to represent the same social constituency.[86] It thus constituted a serious inroad into PNM power and a threat the magnitude of which was shown in a poll conducted by St Augustine Research Associates early in 1981. Asked, 'If an election were to be held in Trinidad and Tobago in the next two months, which party would you vote for?' 29% of a sample of 561 opted for the ONR, 28% for the PNM, 14% for other parties, with the remainder indifferent or uncommitted.[87] Equally promising for the ONR were replies to two other questions. The first, 'Do you think

Trinidad and Tobago: oil capitalism and 'presidential power'   69

Dr Williams should resign as Prime Minister to make room for a successor?' saw 50% answering 'yes' and 35% 'no'. The second took this further, asking, 'If the Prime Minister did in fact resign, whom would you like to see become the next Prime Minister of Trinidad and Tobago?' The replies here were 33% for Hudson-Phillips, 33% 'don't know' and 34% for other candidates or no one.[88] True enough, this poll did not predict victory for the ONR — but then, neither did it for the PNM, and that in itself was a considerable change from the situation only nine months earlier.

As the election year of 1981 loomed Williams, paradoxically, was both stronger than ever before and also weaker. His strengths were his hold on power at the centre — the authoritative command of the decision-making processes of a government growing ever larger as it sought, with some success, to manage the consequences of dependency. His weaknesses were those associated with the performance of this government as it affected the citizenry in their daily lives — bestowing a favour here or apportioning blame there in ways so arbitrary they could be interpreted as either capricious or pernicious according to context and observer. How this situation might be translated into votes on the day no one, not even the most experienced commentator, was prepared to say. What was now apparent, though, and not before, was that Williams himself was preparing to run for office for a sixth consecutive term.

Death, succession and evaluation

On 25 January 1981 Williams ended his address to the Silver Jubilee convention of the PNM as follows:

> So, my dear friends and party colleagues, here we stand after 25 years, to report on our stewardship, to establish our readiness and fitness to continue the struggle on which we embarked 25 years ago, then as now against the rest, then as now with powerful vested interests against us, then as now with the mightiest force in the country in our support, ready to go at the word of command, keeping our powder dry till we see the whites of their eyes, confident in the support of the Lord God of Hosts who will rule, as he has ruled so often in our 25 years. Great is the PNM and it will continue to prevail.[89]

Nine weeks later President Ellis Clarke was to inform the nation he was dead. The immediate reaction, as reported in the press,[90] was one of loss and sorrow at the passing of a great man. But if his death constituted a shock for the country, it did not create a trauma. Contrary to expectations, the succession was accomplished with a minimum of fuss, President Clarke naming one of the deputy leaders of the PNM, George Chambers, as Prime Minister. He was not the obvious choice, though his confirmation in office proceeded without opposition. How this was achieved thus invites consideration of the complex of factors involved. It also points out the nature of the political system Williams left behind.

Most important was the government. In his final speech Williams reiterated a theme he had emphasised earlier when he noted:

> The Prime Minister has remained as virtually chairman of the Cabinet, co-ordinating, establishing priorities, working through committees, the work of which is done by the Ministers without even the presence of the Prime Minister. The result is that, something of which few countries can boast, Trinidad and Tobago now has a team of eight or so top flight speakers and negotiators, ready to deal with any subject in any part of the world, presenting the point of view of the Cabinet as a whole. Pride of place herein is the number of Ministers in the Ministry of Finance, all given increasing insight into our financial problems and ramifications.[91]

Chambers was one of these – and, indeed, had perhaps the greatest experience within the Cabinet of matters directly relating to finance. This, however, was not enough without recognising the system of power Williams established. As noted earlier (and above) it was one where Williams retained ultimate authority while permitting delegated decision-making and limited initiative in important and sometimes sensitive areas. This was supplemented at a second level, outside the Cabinet, by what Parris has accurately described as 'rule through interlocking directorates'.[92] The reference here is to the existence of some 300 persons of ability running the public sector, all of whom were 'political appointees' of the Prime Minister and whose role was 'to communicate the political line which is dominant at the particular time'. At the apex of this small and select constituency, according to Parris, were twenty-two individuals, all men, chiefly of African descent, whose average age was fifty-one. Chambers, of African descent and aged fifty-two, though not one of them, fitted the profile perfectly. He did so by giving legitimacy to technocracy in the form of the party. The PNM thus constituted an indispensable and necessary base for power. Chambers could deliver this, not only because he was close to the party stalwart in the sense of being the self-educated man of lower middle-class origins, but also because he was thought to be loyal to Williams's memory, the recollection being that he was the man who in 1973 had not sought power but, on the contrary, had led a movement to persuade Williams to stay on.[93] Chambers thus represented continuity, as opposed to change for change's sake. As he told the PNM special convention which nominated him unopposed on 9 May, 'What is wrong must be put right, and what is right must stay right',[94] with the emphasis, as his first few months in office showed, on the latter without compromising in any fundamental sense his commitment to the former. Obviously important here were the imperatives of the impending election and a desire not to rock the boat in troubled waters. They are the third factor facilitating a smooth succession and one in which 'chance', as Machiavelli would have it, played as much a part as judgement in determining the final outcome. Without, that is, changing very much, Chambers performed the miracle of transforming the PNM's fortunes at the polls. In one taken in June and published in the *Trinidad Express* in July, he emerged as clear favourite for the national leadership, with 35% as compared to 14% for his nearest rival, Hudson-Phillips. Of those polled, 29·2% said they would vote for the PNM and only 10·9% for the ONR.[95] Chambers, it appeared, had not only learnt from the departed master but was actually in the process of surpassing him – a feat confirmed in the resounding win for the PNM in the general election of 9 November, subsequent analysis of which by an experienced observer drew the comment that 'Trinidad is inexorably moving towards a dominant one-party state'.[96]

Any pronouncement of political stability as Williams's major legacy must be qualified, however, by a frank recognition of the economic foundation on which it rests. From 1974 to 1980 he presided over a boom in the economy in which real output grew at an average annual rate of 7% and government revenues at an average annual rate of 44%.[97] Much of this, of course, can be ascribed to one factor alone – the continued domination in the economy of the petroleum sector, which in 1980 was to contribute some 35% of GDP, nearly three times the contribution of the next highest sector (transport, storage and communications), nearly five times that of manufacturing excluding petroleum and eighteen times that of agriculture.[98] Acute dependence on this area has been recognised and government-directed policies of diversification have been vigorously promoted to attempt to change it. Pride of place has gone to the establishment of the energy-based industries, to which massive commitments have been made. This has not been without considerable risks, which Williams presumably knew and weighed carefully before taking. 'I understand what Nasser meant, when he showed me proudly his steel plant, by saying, since you have hydrocarbons, go for steel. I understand now what Luxembourg meant, a tiny country in the EEC, with an incredible steel complex, when it told me years ago, go for steel,'[99] was how Williams attempted to rationalise his decision to the people of Trinidad and Tobago. Any deficiency in his judgement here will cast a long shadow over the future of the country.

Notes

The author gratefully acknowledges the assistance of a grant from the British Academy for fieldwork in Trinidad in 1978.

1 P. K. Sutton, 'Dr Eric Williams and politics in Trinidad', in *Caribbean Societies*, Vol. 1 (Collected Seminar Papers, No. 29), Institute of Commonwealth Studies, University of London, 1982, p. 54.
2 See, in particular, S. Ryan, *Race and Nationalism in Trinidad and Tobago*, Toronto, 1972.
3 See I. Oxaal, *Black Intellectuals Come to Power*, Cambridge, Mass., 1968.
4 See K. Bahadoorsingh, *Trinidad Electoral Politics: The Persistence of the Race Factor*, London, 1968.
5 'Insider' accounts confirm this – see C. L. R. James, *Party Politics in the West Indies* (formerly PNM Go Forward), San Juan, n.d.; W. Mahabir, *In and Out of Politics*, Port-of-Spain, 1978; and P. Solomon, *Solomon: an autobiography*, Port-of-Spain, 1981.
6 Lloyd Best, 'Options facing Williams, the ruling party and the country, *Trinidad Express*, 31 May 1969.
7 Sutton, 'Dr Eric Williams and politics in Trinidad' in *Caribbean Societies*, p. 56.
8 See, in particular, M. Kroll, 'Political leadership and administrative communications in new nation states: the case study of Trinidad and Tobago', *Social and Economic Studies*, 16, 1, 1967.
9 See Eric Williams, *Inward Hunger: the Education of a Prime Minister*, London, 1969, p. 322, and PNM, General Council – Research Committee, *The Party in Independence*, Port-of-Spain, 1964.
10 See, in particular, Y. K. Malik, *East Indians in Trinidad: a Study in Minority Politics*, London, 1971.
11 Ryan, *Race and Nationalism in Trinidad and Tobago*, pp. 285–87.
12 See his *Inward Hunger* and *From Columbus to Castro: The History of the Caribbean 1492–1969*, London, 1970.
13 See P. K. Sutton, 'Black Power in Trinidad and Tobago: the crisis of 1970', *Journal of Commonwealth and Comparative Politics*, 21, 2, 1983.
14 People's National Movement, *The Chaguaramas Declaration – Perspectives for the New Society* (Approved at a Special Convention, 27–29 November 1970), Port-of-Spain, n.d.
15 Of those registered to vote only 33% did so. As virtually the sole significant party the PNM won all the seats.

16 See People's National Movement, *Address by the Political Leader Dr Eric Williams* (Fifteenth Annual Convention, September 1973), Port-of-Spain, n.d.
17 *Prime Minister's Eleventh Independence Anniversary Message 1973*, Press Release No. 520, 30 August 1973.
18 Figures from Trinidad and Tobago, *Third Five-Year Plan, 1969–1973*; and Trinidad and Tobago, *Review of the Economy, 1972*.
19 Figures from V. Mulchansingh, 'The oil industry in the economy of Trinidad', *Caribbean Studies*, 11, 1, 1971.
20 Figures from *Third Five-Year Plan*.
21 Trinidad and Tobago, Ministry of Planning and Development, *Causes of Unemployment in Trinidad and Tobago and some Remedial Measures*, mimeo, 1970.
22 I. Jainarain, *Trade and Underdevelopment: A Study of the small Caribbean Countries and large Multinational Corporations*, Institute of Development Studies, University of Guyana, 1976, pp. 332–3.
23 F. Rampersad, 'Overseas investment and fiscal policy in Puerto Rico, Jamaica and Trinidad and Tobago' in P. Ady (ed.), *Private Foreign Investment and the Developing World*, New York, 1971.
24 Dr. E. Williams, 'The purpose of planning' in M. Faber and D. Seers (eds.), *The Crisis in Planning*, Vol. 1, London, 1972, pp. 46–7.
25 See C. Parris, *Capital or Labour? The Decision to Introduce the Industrial Stabilization Act in Trinidad and Tobago*, Working Paper No. 11, Institute of Social and Economic Research, UWI, Jamaica, 1976.
26 Trinidad and Tobago, *White Paper on Public Participation in Industrial and Commercial Activities*, 1972.
27 'Economic dependence: a quantitative interpretation', *Social and Economic Studies*, 22, 1, 1973.
28 *Ibid.*, p. 94.
29 Production was at the following levels: 1968, 67 million barrels; 1971, 47·2 million barrels; 1973, 68·2 million barrels. Figures from Organisation of American States, *The Economic and Social Development of Trinidad and Tobago: Characteristics, Policies and Perspectives*, OEA/Ser. 4/XIV, CEPCIES/99, October, 1975.
30 Trinidad and Tobago, Central Statistical Office, *Annual Statistical Digest, 1978*.
31 For details see S. Ryan, *The Politics of Succession: A Study of Parties and Politics in Trinidad and Tobago*, mimeo, UWI, Trinidad, chapter 1.
32 See Sutton, 'Dr Eric Williams and politics in Trinidad' in *Caribbean Societies*.
33 'Prime Minister's Independence Day Message 1974' in Dr. Eric Williams, *Forged from the Love of Liberty* (selected speeches of Dr Williams compiled by P. K. Sutton), Port-of-Spain, 1981, p. 80.
34 *Ibid.*, p. 76.
35 'Economic Transformation and the Role of the PNM', *Address* by the Political Leader, Dr Eric Williams, to the sixteenth Annual Convention of the PNM, October 1974, mimeo.
36 See 'The Energy Crisis – 1975', *An Address to the Nation*, at Point Fortin,

11 April 1975, Trinidad and Tobago, 1975, and 'Speech by Dr Eric Williams' at Harris Promenade, San Fernando, 13 April 1975, mimeo.
37 Figures calculated from Trinidad and Tobago, *Accounting for the Petrodollar*, 1980, table 31.
38 *Ibid.*, pp. 46–8.
39 See '1976 and 1956', *Address* by the Political Leader, Dr Eric Williams, to the Seventeenth Annual Convention of the PNM, October 1975, mimeo.
40 House of Representatives of the Republic of Trinidad and Tobago, *Budget Speech 1978* (Dr E. Williams, 2 December 1977), p. 10.
41 Calculated from Trinidad and Tobago, *Review of Fiscal Measures in the 1980 Budget*.
42 The perils of careless divestment are set out in Williams, 'Address at the Opening of the First Branch of the National Insurance Board, 19 May 1977' in his *Forged from the Love of Liberty*.
43 Cited in Ryan, *Race and Nationalism in Trinidad and Tobago*.
44 Constitution Commission of Trinidad and Tobago, *Report of the Constitution Commission*, para. 51.
45 See, in particular, D. Moore, 'The Westminster model under attack: the report of the Constitution Commission of Trinidad and Tobago', *Journal of Constitutional and Parliamentary Studies*, 8, 3, 1974.
46 'Constitution Reform', Speech by the Prime Minister in the House of Representatives, 13 and 17 December 1974, Trinidad, 1975, p. 82.
47 'PR: To Dissolve the Present PNM Majorities', *Address* by Dr Williams to the Sixteen Southern Constituencies Rally, 1 April 1973, Port-of-Spain, n.d.
48 See 'Constitution (Republic) Bill', Second Reading, House of Representatives of Trinidad and Tobago, *Hansard*, Speech by Dr Williams, cols. 677–734.
49 'Economic Transformation and the Role of the PNM', mimeo, pp. 29–30.
50 '1976 and 1956', in Williams, *Forged from the Love of Liberty*, Part 2, Section 4, p. 185.
51 '1976 and 1956'; see also Speech on the Constitution (Republic) Bill, cols. 685–730.
52 '1976 and 1956', in Williams, *Forged from the Love of Liberty*, p. 191.
53 *Trinidad Guardian*, 26 May 1976.
54 See his 'Address to the Fourteenth Annual Convention of the PNM, September 1972' in Williams, *Forged from the Love of Liberty*, Part 2, Section 1; and his 'Address to the Fifteenth Annual Convention of the PNM', September 1973.
55 *Ibid.*
56 Letter from Karl Hudson-Phillips to the Prime Minister in *Trinidad Guardian*, 30 May 1976.
57 *Tapia*, 3 October 1976.
58 See Ryan, *The Politics of Succession*, chapter 5.
59 S. Ryan, E. Greene and J. Harewood, *The Confused Electorate – A Study of Political Attitudes and Opinions in Trinidad and Tobago*, St Augustine, Trinidad, December 1979, table 1.37, p. 23, and table 1.40, p. 25. The figures for others were: Panday, 7%; Robinson, 6%; Best, 5%;

others, 10%. Significantly, 37% said they didn't know.
60 'Problems of Industrialization', *Address* by Dr E. Williams to the Twentieth Annual Convention of the PNM, 29 September–1 October 1978, mimeo, p. 10.
61 House of Representatives of the Republic of Trinidad and Tobago, *Budget Speech 1980* (Dr E. Williams, 30 November 1979), pp. 3–4.
62 House of Representatives of the Republic of Trinidad and Tobago, *Budget Speech 1979* (Dr E. Williams, 1 December 1978), p. 8.
63 'Twenty-fifth Anniversary', Political Leader's *Address* to the Special Convention of the PNM, 25 January 1981, mimeo, paras. 75–6.
64 Calculated from *Review of Fiscal Measures in the 1980 Budget*, pp. 11–12.
65 *Ibid.*
66 Calculated from *Accounting for the Petrodollar*, table V.
67 R. Ramsaran, *The Growth and Pattern of Public Expenditure in Trinidad and Tobago 1959–1979*, mimeo, p. 103.
68 'The PNM in the Next Five Years, 1976–1981', *Address* by Dr E. Williams to the Eighteenth Annual Convention of the PNM, 3 December 1976, mimeo, p. 63.
69 'The Caribbean Man', *Address* by Dr E. Williams to the Twenty-first Annual Convention of the PNM, 29 September 1979, mimeo, pp. 3–4.
70 See *Caribbean Contact*, June 1978.
71 'The Party's Stewardship, 1956 to 1980', Speech to the Twenty-second Annual Convention of the PNM, 26 September 1980, in Williams, *Forged from the Love of Liberty*, p. 425.
72 *Latin America*, Caribbean Regional Report, 28 March 1980, p. 6.
75 'Corporation Sole' was designated in 1973 as the device by which the Minister of Finance held all government shareholdings by virtue of his office.
74 Calculated from Trinidad and Tobago, *Accounting for the Petrodollar*, table 31.
75 *Ibid.*, pp. 48–50; *Budget Speech 1978*, p. 61; and House of Representatives of the Republic of Trinidad and Tobago, *Budget Speech 1981* (Dr E. Williams, 5 December 1980), p. 49.
76 G. Chambers, *Budget Speech 1982*, mimeo, p. 64.
77 Ryan, *The Politics of Succession*, p. 204.
78 See, in particular, the speeches of Robinson and Selwyn Richardson in J. G. Davidson (ed.), *Tobago versus PNM*, Port-of-Spain, 1979, pp. 11–67.
79 *Caribbean Contact*, December 1980.
80 *Report of the Constitution Commission*, para. 235.
81 For details see W. Richard Jacobs, 'Patterns of Political Corruption in Caribbean Society: A Comparative Study of Grenada, Jamaica and Trinidad and Tobago', *Occasional Papers*, ISER, UWI, St Augustine, Trinidad, December 1978, pp. 49–91.
82 *Budget Speech 1981*, p. 35; 'Twenty-fifth Anniversary' *Address* (PNM), para, 137.
83 'The Caribbean Man', pp. 82–3.
84 Of the 250 PNM party groups which made valid nominations, 224

　　　 endorsed Hudson-Phillips as against twenty-six for K. Mohammed. Ryan, *The Politics of Succession*, p. 20.
85　In the elections the PNM won control over every county and municipal council in Trinidad, conceding only eleven of the 113 seats available, and winning thirty-one of them unopposed. See *Caribbean Contact*, May 1980.
86　See, in particular, Selwyn Ryan, 'The Church that Williams built – electoral possibilities in Trinidad and Tobago', *Caribbean Review*, 10, 2, 1981.
87　*Ibid.*, table 4, p. 45.
88　*Ibid.*, tables 1 and 2, p. 45.
89　'Twenty-fifth Anniversary' *Address* (PNM), para. 139.
90　See, in particular, the commemorative issue of *Caribbean Contact*, May 1981.
91　'Twenty-fifth Anniversary' *Address* (PNM), para. 138.
92　Carl Parris, 'Power and rule under Williams', *Trinidad and Tobago Review*, 5, 12, 1982. This is not the first reference to interlocking directorships. A. N. R. Robinson in his *The Mechanics of Independence: Patterns of Political and Economic Transformation in Trinidad and Tobago*, Cambridge, Mass., 1971, notes in respect of difficulties surrounding the passage of the 1966 Finance Act, 'The defects of the corporate structure had other fiscal implications; the social and political consequences were tight control of the economy of the country by an oligarchy made all the more limited through interlocking directorships,' p. 92. Is the situation Parris refers to one of private economy absorbing public polity, or public economy absorbing private polity? Or is it just state capitalism?
93　Of the forty PNM party groups who refused to nominate a candidate to succeed Dr Williams, twenty were from the St Anns constituency represented by Chambers.
94　*The Advocate-News*, Barbados, 11 August 1981.
95　See *Caribbean Contact*, September 1981.
96　John La Guerre, 'The General Elections of 1981 in Trinidad and Tobago', Department of Government, UWI, St Augustine, Trinidad, February 1982, mimeo, p. 1.
97　Figures from Chambers, *Budget Speech 1982*, pp. 9 and 11.
98　Figures from 'Trinidad and Tobago Country Notes' in United Nations, Economic Commission for Latin America, *Economic Activity in 1980 in Caribbean Countries*, CEPAL/CARIB 81/10, Appendix, table 2.
99　'The Caribbean Man', p. 78.

Clive Y. Thomas

3 Guyana: the rise and fall of 'co-operative socialism'

At Guyana's independence in 1966 the dominant structural relations of the economy, and the social form through which they were systematically reproduced, portrayed many of the classic features of underdevelopment–dependency relations. In addition to the small size of the economy, the following conditions prevailed:

1. Primary producing economic structures, predominantly rural, with agricultural resources organised within a backward agrarian system, cultivating crops principally for export to overseas metropolitan markets. The principal commodities produced were sugar, rice and bauxite-alumina. Except for a small fraction of sugar cane cultivation (less than 10%), sugar was grown and processed by two foreign-owned plantations.[1] Sugar was the dominant crop, accounting for the largest share of value added, employment, foreign exchange earnings, capital accumulation, crop land and agricultural infrastructural resources.[2] Rice, initially cultivated as a domestic staple, had grown into a significant export cash crop, with most sales to the Caribbean market. Rice cultivation was not, however, a significant competitor to sugar for agricultural resources. Indeed, the industry was developed as a complement to sugar under the patronage of both the colonial state and the plantations.[3] Bauxite-alumina production took place in two enclave mining areas under the control of two aluminium-producing transnational corporations (TNCs) – Alcan, based in Canada, and Reynolds, based in the USA. The bulk of production was under the control of the Canadian corporation.

2. The dominant role of foreign ownership and control of the country's productive assets was paralleled by the dominant role of foreign financing of 'development', foreign control of technology, extensive reliance on foreign trade with the countries from which the

external owners and controllers of the productive assets originated (the UK and North America), the decisive role of the foreign element in economic decision-making, and foreign control of the financial mechanisms. All these facilitated the heavy drain overseas of the surpluses internally appropriated within these economic structures.

3. Export specialisation in primary products was complemented by extensive dependence on the importation of foodstuffs (although the economy was principally agricultural), intermediate goods (particularly oil and fertilisers), and capital equipment. This highlighted the lack of internal differentiation of the economy and the absence of any significant backward and forward inter-industrial linkages.

4. As it was organised the economy was unable to provide enough jobs for the workforce. While accurate data are hard to come by, it is not an overestimate to say that about a quarter of the workforce was unemployed, while a large proportion of the employed were underemployed. For example, the sugar workers worked only seasonally.

5. In view of the above, the economy displayed widespread signs of poverty: poor housing, underdeveloped medical services, inadequate educational services, backward communications facilities, a dearth of recreational facilities, and so on.

During the 1960s, in particular, these structures yielded impressive growth rates. These have generally continued during the 1970s, as Table 3.1 illustrates, but because internal accumulation is profit-oriented and heavily dependent on world market conditions they have also been subject to considerable secular as well as cyclical fluctuations. One might be tempted to say that these starkly dependent economic structures mirrored the colonial status of the country at the time, but, as we shall observe, a decade and half after independence, and a decade after the establishment of 'co-operative socialism', these structural relations remain essentially the same.

The system of economic reproduction into which these structures are integrated cannot, of course, be properly understood in isolation from the long historical process by which the capitalist mode of production was internationalised after the sixteenth century. In Guyana this process resulted in the development of two important sets of divergences in the system of reproduction of the material conditions of social life.[4] The first of these is the divergence between the pattern of resource use and the demand structure as it exists, and as it is likely to develop as commodity relations become more generalised and

Table 3.1 Sectoral contribution to GDP at factor cost (%), 1970–80 (current prices)

Product	1970	1971	1972	1973	1974	1975	1976	1977	1978	1979	1980
1. Sugar	12·4	14·8	14·5	11·7	29·0	30·6	20·4	10·2	14·5	14·2	14·8
2. Rice	3·7	2·9	2·1	2·8	3·4	3·8	2·9	5·7	4·4	3·4	3·8
3. Livestock	2·2	2·3	2·2	2·3	1·7	1·8	2·1	2·6	2·7	3·2	3·1
4. Other agriculture	2·9	2·9	2·9	3·1	2·2	2·0	2·4	3·5	3·4	3·6	3·6
5. Agricultural sector (1+2+3+4)	21·2	23·0	21·7	19·9	36·3	38·2	27·8	22·0	25·0	24·4	25·3
6. Fishing	1·1	1·1	1·1	1·2	1·2	0·9	0·9	1·3	1·2	1·3	1·3
7. Forestry	1·1	1·0	1·1	1·0	0·9	0·8	0·9	1·0	1·0	1·0	1·2
8. Manufacturing	8·2	7·9	7·9	7·6	5·8	6·3	7·4	8·5	7·6	8·3	8·1
9. Mining and quarrying	20·4	18·3	17·0	14·0	13·5	13·0	13·7	16·1	15·7	13·3	16·5
10. Transport and communication	5·9	6·0	6·2	6·4	5·1	4·6	5·7	5·7	5·7	6·2	5·6
11. Engineering and construction	7·9	7·8	8·1	8·1	6·2	6·2	7·6	7·5	6·6	7·3	6·7
12. Distribution	11·5	11·0	11·1	11·2	8·9	8·7	9·9	9·8	9·2	9·1	8·6
13. Services and rent	9·6	9·9	9·6	9·7	7·4	6·8	8·0	8·4	7·8	7·9	7·9
14. Government	13·2	14·1	16·3	21·0	14·9	14·6	18·0	19·7	20·3	20·1	18·7
15. % Change in total product on previous year	+6·7	+5·9	+6·9	+8·9	+50·0	+26·8	−3·6	−0·5	+1·1	+2·6	+1·5

Notes.
(a) The agricultural sector is given as the sum of rows 1–4. In the Guyana national accounts data sugar and rice processing are placed in the manufacturing sector. For our purposes, because a very primary state of processing is involved, they are treated as agriculture. The manufacturing sector (row 8) therefore includes manufacturing other than rice and sugar processing.
(b) The data for 1980 are not yet final.
Source. Ministry of Economic Development, Statistical Bureau, and Bank of Guyana Annual Reports.

entrenched in the national economy. The reason for this is that European conquest and settlement of the country resulted first in the overthrow of the communal and other pre-capitalist production relations which dominated, and thereafter systematically garnered domestic resources for the servicing of overseas markets in Europe and elsewhere. The development of tropical agriculture under slavery, enclave mineral production for export, etc., all reflect this orientation. The consequence is that local demand has been and continues to be by and large satisfied by imports. The export sector, as we have already noted, has become highly specialised in the production of a small number of commodities and confines the sale of its output to one or two major capitalist markets. The second divergence is that, as commodity relations have developed, ability to purchase in the market has become the principal determinant of consumption. Given the widespread unemployment and poverty which have historically prevailed, the result has been that since colonial times the needs of the broad masses of the population have not influenced production in any significant way. Except for residual subsistence sectors, commodities available from both local production and imports are allocated on the basis of purchasing power.

Around this dynamic of reproduction of the material needs of the society as a whole has grown up a complex of political, social and legal relations (patterns of ownership of productive forces); institutions and other decision-making structures and organisations; and ideas and ideologies.[5]

Criteria for evaluation: development for whom?[6]

If the proposition is that the above structures and system of economic reproduction conform to the classic pattern of dependency/ underdevelopment relations, an alternative conception of social relations should logically provide criteria for evaluation of development and guide the critical thrust of this chapter. An explicit but brief presentation of this is attempted in this section.

A good starting point is the insight afforded by the systematic divergences referred to above. First, in direct contrast to what has prevailed, and irrespective of the behaviour of *per capita* product, development occurs when production is oriented towards satisfying the basic needs of the masses. Here, 'the masses' refers to all those 'who do not have any power in society derived from property, wealth, religion, caste, expertise or other such sources not widely shared'.[7] To these minority attributes I would add political party affiliation as being particularly applicable, since in these countries as the state becomes more and more the exclusive domain of a narrow group, and undemocratic, non-representational politics become the rule, political loyalty or affiliation to the ruling class becomes vital to the enhancement of one's status. This applies as much to the lowly peasant looking for land, markets or credit as to capitalists looking for import licences and trading permits. In these circumstances production for the basic needs of the masses implies at the very least a systematic, conscious attack on the widespread poverty which prevails. Development implies, therefore, that the elimination of poverty is not treated as an incidental consequence of production, as is the case when profit is the sole determinant. The needs referred to here are personal (food, clothing, housing) and public/collective (health, sanitation, education, culture, recreation, etc.). They are also, as may be readily inferred, both material and non-material.

Second, development also means that the satisfaction of these basic needs should be generated through the effective exercise of the right to work. By this is meant not only that all those who want jobs should have them but also:

1. The right to a job without coercion as to place or type of job, given the particular skills.
2. A framework of industrial relations which permits free collective bargaining and effective (as distinct from nominal) representation within bargaining units.
3. A work process that allows for effective worker involvement and control.
4. The protection of health and guarantee of education and training for all workers for the tasks they are engaged in.

The objective here is to situate work in a 'self-realisation process' where it becomes both an *end* and a *means* of development.

Third, in view of the historical formation of underdevelopment and the resultant pattern of producing what is not consumed locally and consuming what is not produced locally, combined with foreign ownership and domination of national production, the reproduction of the material conditions of life should be based on self-reliant and indigenous patterns of growth. This is the only just and, in the long run, sustainable pattern of growth, as it is premised on the explicit objective of reversing dependency relations and situating development in the context of the capacity of the people of the country to develop themselves.

Development also requires that the work process should be based on the democratisation of power in society combined with the effective, as distinct from nominal, exercise of fundamental rights, that is, the rights to free and fair elections, free expression and organisation, respect for the rights of the individual, protection from repression and torture, and so on. This democratisation of power also means the democratisation of the decision-making structures of society, and two requirements for achieving this are an equitable distribution of wealth and income, and an equitable access to the use and management of society's resources. Such a democratisation of power is not possible without equitable access to information, as the effective possession of information is necessary if sound decisions are to be made and control from below, or 'popular participation' in the process of development, as it is sometimes termed, is to be efficient, while at the same time reflecting the requirements of the largest possible majority.

Finally, development is conceived in this chapter as implying a preservation of environmental stability and a halt to the degradation of the environment which has so far accompanied the growth of national production. This conception is situated in an unequivocal acceptance of the global and universal responsibilities of all societies to sustain life on planet earth.

It should be obvious that the intent here is to ensure that the material and other needs referred to are treated dynamically. That is, their interaction and historical expansion are implied in the formulation. What is perhaps not so self-evident, and what should be made clear, is that this conception of development does not mean advocacy of a mechanical route to salvation which all countries are bound to pursue. On the contrary, needs, and the methods of determining them, are taken to be always historically and culturally specific. It is for the people of each society, *in their collectivity*, to determine the ways in which this conception of development is embodied in their material production and social arrangements. Any other approach would be elitist and vanguard determinist. It would also be socially and politically anti-democratic, and as such would not enhance the liberation and development of people, which ultimately emerges as the fundamental objective of development. Development, therefore, means the progressive de-alienation of man from the process of social reproduction.[8]

The colonial view of development in Guyana

The principal predecessor to the economics of 'co-operative socialism' in Guyana has been twentieth-century colonial economics as practised and preached by the colonial authorities. This embraced the view that development was best measured by the rate of increase of *per capita* national product, with the stipulation that even such a narrowly conceived notion of development was impossible in Guyana in any truly long-term sustainable form because of the size, geographical environment and cultural and racial characteristics of the people. This view was firmly adhered to, despite the attraction to the wealth of the country on which the first wave of conquest and colonisation was based, Guyana being the supposed site of Raleigh's El Dorado and its lush tropical rain forest a sign of the unlimited fertility of the soil! Racial theories, as well as geographically determinist views of development, were used to explain why poverty and unemployment would endure, and why capitalism would not, indeed could not, develop in Guyana as it did in Europe and North America. Arising out of this, cetain policies of economic management were introduced and promoted by the state. Among them were:

1. *Agricultural production.* In Guyana the rural structure combined a highly concentrated large-scale plantation export-producing sector with a large number of small and landless peasants engaged in subsistence agricultural production for the domestic market and in some cases for export also (rice and sugar). The fact that agricultural production dominated economic activity was taken to be an accurate reflection of the natural resource endowment and comparative advantages of the country. Consequently much attention was paid to the promotion of agricultural production. A variety of measures were introduced, among the most important being land settlement schemes. These were promoted because it was expected that they would alleviate population pressures on the land, while catering to some of the peasant discontent which fuelled the struggles for independence during this century. Secondly, in order to ensure colonial state regulation over the use of the land and the distribution of its product, peasant production was linked wherever possible either to state marketing agencies or to co-operative marketing arrangements. Thirdly, technical services in the form of agricultural credit, extension services, etc., were provided as aids to agriculture. And last but not least, because plantation agriculture generally predominated, great efforts were made to

rehabilitate, modernise and reorganise these production units wherever possible. The means by which these aims were pursued was providing subsidies to the plantation owners in order to facilitate the recomposition of their capital stock; pursuing wage and employment policies designed to stabilise the flow of labour to the plantation economy; providing basic infrastructural facilities to encourage production, and providing colonial preferential trading arrangements.

2. *Infrastructural production.* During this period efforts were also made to develop, through state financing, an adequate infrastructure for private production. These encompassed land reclamation schemes and drainage and irrigation works, as well as the usual public utilities: electricity, sanitation, pure water supply, etc.

3. *Foreign capital inflows.* Consistent with the colonial view of the 'developmental capacity' of the territory was the insistence that whatever development was possible depended on the continued inflow of foreign finance, skills and technology, whether on a private, state or international agency basis. General efforts were made to encourage this through maintaining the effective unification of the territory with the metropolitan market of the UK. Thus currency and banking arrangements, e.g. Currency Boards and exchange rate regimes, were premised on the effective unification of individual capital markets of the region with that of the United Kingdom.

4. *Population control.* For the Empire as a whole and for the West Indies in particular, non-whites were held to be reproductively prolific. Despite the slave trade and indentured immigration it was felt that this was responsible for the territories of the region having become overpopulated. In order to restore a more favourable balance to the people:land ratio, family planning measures were introduced. In Guyana the gross people:land ratio has always been very low. However, as most of the population has settled on the narrow and relatively fertile coastal strip, intense pressure on land resources has always existed.

The overall objective of these policies was to promote the reproduction of the classical colonial division of labour. One important aspect of this was that industrial development of any real magnitude was ruled out. The main promoters and hence beneficiaries of economic progress were taken to be the large-scale plantation owners. Efforts to improve the lot of the peasantry were politically motivated by the need to contain the mass movement for independence which at the time was being fuelled by the acute land shortage which the peasants faced and the consequent increase in destitution, unemployment and large-scale migration to the cities. The resultant efforts were, however, far from adequate; and over the period peasant discontent was intensified by the continued privileged status of plantation agriculture. Migration to the cities also resulted in the creation of massive open unemployment and turned the urban centres into seething cauldrons of discontent. Riots, insurrections and other populist outbursts became more and more a part of the agitation for national independence. Since the colonial policy of national subjugation favoured almost exclusively the backward plantocracy and the Colonial Office officials who managed the colonial state machinery, it stood in direct opposition to the interests of the broadest sections of the people, thereby strengthening the basis for an alliance of all classes and strata around the agitation for national independence. In the last analysis, however, these colonial policies were rapidly becoming obsolete, as they sought to perpetuate in the context of a rapidly changing imperialist world system the relatively primitive conditions of the early colonial forms of exploitation.

The early nationalist phase of development

While twentieth-century colonial economics was the principal antecedent to the political economy of co-operative socialism, the replacement of the former by the latter was preceded by an early nationalist phase of development which it is important to understand in order to appreciate some of the premises of co-operative socialism.

The disruptions to the world economy occasioned by the second world war and the subsequent collapse of certain agricultural prices in the 1950s and 1960s led to great internal upheavals in the rural economy of the entire Caribbean. These developments coincided with the political decline of the plantocracy in national politics and the rise in importance of petty-bourgeois elements in the state machinery and national economy which accompanied the movement to universal adult suffrage and self-government. The political impetus for these developments was of course agitation for national independence. This reached a high point in Guyana in 1953, when, in the first elections based on universal adult suffrage, the first Marxist government in the British Empire was elected to office. Cheddi Jagan at the head of the People's Progressive Party (PPP) formed this government. However, after 133 days the British intervened, suspended the constitution and returned the country to 'Crown colony' status. This reversal ensured that the working-class–peasant orientation which the PPP promised to bring to bear on economic policy could never materialise. A combination of 'interim' colonially appointed governments and a limited self-governing constitution was henceforth to be the status of the new state, until independence in 1966.

Despite this reversal, there continued to be ideological resistance to the colonial view that industrialisation was impossible. It was fed by the continued agitation for independence, combined with the pressures of increasing poverty and unemployment, aggravated by the rapid mechanisation of the estates as they sought to reduce costs and minimise their reliance on a workforce increasingly influenced by developing trade unionism – and by the national independence movement itself. In particular, sections of the emerging petty bourgeoisie soon recognised that the development of some form of industry was the only way to secure the material conditions for their own self-reproduction and eventual growth into a national bourgeoisie. They began to champion the cause of industrialisation, and the model then being developed in Puerto Rico was pointed to as proof of its real possibilities in the region. The particular advantage of the Puerto Rican example to this class was its supposed vindication of the possibility of industrialisation in a small country, predominantly peopled by persons of non-Anglo-Saxon descent. Although independence did not come until 1966, the influence of this 'nationalist'-oriented phase of colonial development policies is evident.

What were the main features of the Puerto Rican model of industrialisation? First, it required state provision of incentives to TNCs to encourage the establishment of branch plants in order to provide jobs, transfer technology and create local markets for their inputs, while also creating opportunities for nationals to acquire business and managerial skills. Incentives varied from fiscal rebates to the construction of industrial infrastructure – industrial estates, harbours, airports, etc. – and the provision of industrial credit to stimulate local small industries. Of special importance also was the minimisation of foreign exchange restrictions on international transactions.

The second requirement was political and industrial peace at all costs, as it was felt that 'over-militant' trade unionism would deter the establishment of branch plants. The contradiction here was that the petty bourgeoisie needed an alliance with the workers and peasants to promote the independence struggles which they led. In the pre-independence period, therefore, the conflict with the working class implicit in this policy was relatively muted, manifesting itself strikingly after independence when the emergent ruling classes had secured control of the state. In this process ideological support for private property played a significant role. State intervention in the economy was promoted as being limited to a supporting role for private initiatives. Workers were exhorted not to strike, such action being propagandised as anti-national, disruptive of the economy and certain to scare away foreign capital, in the end making the workers worse off. The 'cold war' conditions of the period favoured a strong anti-communism and anti-socialism in this propaganda.

A third aspect of the dependent industrialisation model was the relative downgrading of agriculture in economic development. In some instances people were actually encouraged to leave the land in the hope that this would increase the productivity of those remaining and so raise rural incomes. This overall reduction of emphasis on agriculture, however, was combined with emphasis on certain sectors, the most important being the large estate-plantation export sector, and certain areas of domestic production, e.g. fruit and vegetables for the tourist industry.

Finally, all the above policy elements required complementary developments in the establishment, where absent, and the improvement, where necessary, of the economic regulatory and management agencies of the state, e.g. central banks, legislation governing foreign exchange transactions, updating of labour legislation, the establishment of employment exchanges, and improvement in public utilities.

None of the measures should be seen in isolation, as they were all integral elements of the thrust to create a process of industrialisation based on the expansion of the activities of the TNCs. This path of development conflicted with the previous colonial one to the extent that it represented the pursuit of the option available to the local petty bourgeoisie for implanting itself in the system of economic reproduction, and so, it was hoped, developing its own variant of the capitalist model.

The pursuit of this path in Guyana reflected above all a new balance of class forces that was being established. The national independence movement eventually brought to an end formal colonial rule and formal domination of the state by Colonial Office officials. While the main actors were the small incipient working class and the peasantry, under the leadership of certain petty-bourgeois elements which had grown up in the colonial state machinery, the professions and local business, the transfer of power to the local state was *premised on the effective exclusion of the mass of workers and peasants from political power.* Control of the state machinery was placed in the hands of the petty bourgeoisie, which was then able to inaugurate the process of dependent industrialisation. To its great surprise and chagrin this path of development did not lead to its effective consolidation; a new strategy had to be evolved, and, given the limited social base of this group, the state was to be the main instrument of its execution. Before we examine this phase of state capitalism it is essential to indicate some major defects of this phase of dependent industrialisation.

Several negative features emerged. The vast majority of the industries that were established were branch plants of TNCs which assembled prefabricated imported inputs for the urban high-income markets. This final assembly 'screwdriver technology' type of industry has a very low domestic value added while employing highly capital-intensive methods. With a small domestic market the production units did not nearly approach their full output capacity. Costs of production were therefore relatively high, despite considerable capital subsidisation. These, however, were passed on to the consumer in the form of higher prices and poorer quality, owing to the monopoly status these firms enjoyed. Profits were uniformly super-normal, and, because of free capital mobility, the external drain of foreign exchange was considerable. Such industries created negligible inter-sectoral links in the national economy, so that the development of the manufacturing sector could not generate any autonomous, self-reinforcing sequences of growth – the correlate of this being the high import-intensity of this sector. Finally, there was the strategic consideration that the TNCs had no further goal other than capturing the local, or in some instances the regional, market, and consequently this sector has developed no real export capability.

The urban concentration of these industries encouraged the flow of people from the rural areas to the cities and, with the disintegration of traditional agriculture, forced the underemployed in the countryside into the open. The capital city became distinguished by its barrios, ghettos and the large number of young people who had never worked. 'Fortunately', it was claimed, the earlier periods of this phase of industrialisation coincided with high rates of external migration, which reduced the pressures somewhat. In the later periods, as stricter quotas were enforced by the receiving countries, emigration was considerably diminished. Today it is only the relatively skilled who are able to obtain residence visas, and this, combined with the flight of professionals out of the country, has led to a serious 'brain drain'.

Co-operative socialism: salient features

The negative features of the Puerto Rican model of development soon led to generalised socio-economic crisis not only in Guyana but in most of the English-speaking Caribbean. In the light of the class interests of the petty-bourgeois strata which controlled the post-independence state, two options were effectively open. The first was to interpret the deficiencies of the model in the failure to pursue it consistently and rigorously. In some of the territories, e.g. Shearer's Jamaica, Barrow's Barbados and to a lesser extent Williams's Trinidad and Tobago, this was opted for, and the 'private enterprise' solution of closer and closer integration into the TNCs' production structure was assiduously pursued. Where, as in Guyana, popular resistance to increasing unemployment, poverty and foreign domination of the economy has been particularly strong the ruling class, led by President Burnham and the People's National Congress (PNC), has opted for the development of state capitalist forms as a counter to TNC branch plant domination. This latter development was signalled by ideological proclamations by the leadership that the country was 'going socialist'. Domestic labels were, of course, put on this development (e.g. co-operative socialism), but they were all designed to show that the state was pursuing its own independent and indigenous path to socialism.

In Guyana the proof of socialist intent was supposedly enshrined in four major policy initiatives of the state: the nationalisation of foreign property in order to assert more national control over the economy; the proclamation that economic policies were henceforth to be geared exclusively towards 'feeding, clothing and housing' (FCH) the nation; the announcement that the dominant form of ownership and social organisation of production would be the co-operative; and a declaration to the effect that the ruling party was 'paramount' over all other parties and the state machinery itself.[9] Each will be examined briefly in turn.

After declaring Guyana a Co-operative Socialist Republic in 1970, the government nationalised a considerable number of industries, so that by the mid-1970s the main producing sectors (sugar, bauxite-alumina), the import trade (through the formation of an External Trade Bureau) and significant sections of the distribution, communications and public transport sectors were state-owned. Thus when the sugar industry was fully nationalised in 1976, the government claimed that it now 'owned and controlled 80% of the economy'. The implementation of these measures was preceded by other efforts to localise the economy and reduce foreign domination. Although on the government's own admission these earlier efforts failed, it is instructive to note that it had already indicated that the objective was to reduce 'foreign' capital domination and not the domination of capital *per se*.

Two of the earlier efforts exemplify this point. The first was the policy of localising the management of the TNCs. It was believed that if this were done the operations of the TNCs would be subject to local control. In this formulation local control was a logical extension of the petty-bourgeois view that the object of the independence struggle was the transfer of colonial state power into their hands, and not into the hands of the masses. Although there was nothing anti-capitalist in this policy, it nevertheless failed. It failed, however, not in the sense that no gains were made in getting local people into the upper echelons of management, but that there was no noticeable effect on the workings of the TNC. In the first place, those who moved into executive positions rapidly became socialised to the ethic of the TNC. Rather than leading to more local control, the effect was to 'delocalise' these managerial elements. Further, the policy underestimated the capacity of the TNCs (aided by modern communications) to substitute positions within their structure. Thus, if there was pressure to bring local people into top management, the amount of local decision-making was reduced by transferring some of it to a regional or head office, relying on the speed of modern communications for transmitting decisions. Finally, the tendency was generally to accede to pressure to localise management in fields which were most visible, e.g. personnel management, public relations. In this way the local elements of the managerial structures became, as it were, buffers for the TNCs in the face of aggressive nationalism.

A second example of this type of policy was the pressure on the corporations to 'go local' by issuing shares on the local market. There is no need to detail the many ways in which they were able to minimise the effects of any local share ownership. Suffice it to say that there has been not a single case of foreign control being diluted by this means. Even if there had been, the effect would again have been to substitute local for foreign capital, rather than to reduce the overall importance of capital in the economy.

Guyana: 'co-operative socialism'

The fact that such policies preceded nationalisation indicates the limited perspectives from which nationalisation was offered as a solution to the developmental *impasse*. It is hardly surprising that it took on the appearance of a 'mortgage finance' transaction. To begin with, expropriation was never attempted; instead the state paid a commercial price for the foreign-owned property. This price has been described as 'mutually agreed upon', and there is no known case where the companies have felt the financial terms to be less than satisfactory. The effect of such purchase has been to convert the assets previously held by the transnational into a foreign debt obligation of the state. Under public pressure the government managed to insert clauses stipulating that repayments would be made out of profits. But here again no case is known of the government refusing to repay a debt because of operating losses. The reason is that after nationalisation these corporations continued to rely on foreign finance (suppliers' credit, bank loans, etc.), which would not be forthcoming if it was felt that a deliberate default was in question. More important, however, nationalisation was accompanied by contracts in the areas of technology, marketing and management. The net effect has been no significant reduction in foreign control over the nationalised enterprises.[10]

The second aspect of co-operative socialism was the FCH programme, which has been more propagandistic than anything else. The 1972–76 development programme which highlighted this new policy appeared as a draft in the middle of 1973 and has never been revised or added to since in any public document. As Standing has observed in his ILO study of poverty and basic needs in Guyana, 'Since the 1972–1976 Development Plan has been "rolled over" into 1977 and since that Plan was often vague on how the Government's explicit objectives were to be met it is not surprising that the country has not made nearly as much progress as envisaged in the original Plan.'[11] He goes on to speak of the fact that 'as far as nutrition is concerned the evidence points to a widespread incidence of malnutrition, which is both cause and effect of ill health'.[12] On housing he points out: 'The Government's 1972–76 Plan aimed to "house the nation" within the five-year Plan period ... once again the rhetorical commitment was not matched by adequate planning.'[13]

The third feature of co-operative socialism was that given a tri-sectoral national economic structure – private, state and co-operative – the co-operative sector should be developed into the dominant one. Co-operative ownership and control would then be the foundations on which socialism is built. Historically, the co-operative sector in Guyana has been and remains a very small part of the national economy. But even here the formal size of the sector still overstates the social significance of co-operative ownership and control. This can be seen both in co-operatives promoted by the state and in those due to private initiative. In the former, e.g. the Guyana National Co-operative Bank, the co-operatives are in effect state-run institutions which operate on ordinary commercial principles, despite the formal co-operative ownership structure. In the private sector, particularly among the economically significant co-operatives, e.g. in the field of construction, they operate on private capitalist principles. Thus many of the co-operatives employ wage labour, as membership does not lead to automatic enfranchisement. In such circumstances the real owners proceed to accumulate on the basis of exploited wage labour. It is an attractive procedure, since co-operatives enjoy tax concessions, making them in this sense a convenient medium for private accumulation. Another instance of this type of distortion is co-operatives formed with specific and limited objectives in mind. A land co-operative may be formed in order to acquire a piece of state land (the regulations for acquisition may make this necessary); after the land has been acquired it is immediately subdivided and exploited on an individual basis.

Finally, the policy of 'paramountcy' is an assertion by President Burnham and his party of their supremacy over the government, other arms of the state and all other political, social and cultural organs. Given the fact that the government did not come to power as a result of either free and fair elections or a popular social revolution, such an assertion was in effect a claim to dictatorship. Despite this, the nationalisations were misread by the main opposition party at the time (Jagan's People's Progressive Party) as a 'turn to the left by the PNC' and the government's pursuit of a 'non-capitalist path' of revolutionary democracy. In 1975 the PPP announced a policy of critical support for the government, even though two years before it had called for 'non-co-operation and civil resistance' after massive electoral frauds in the 1973 elections when, with the aid of the army, the government had seized the ballot boxes and frustrated the will of electorate. There is little doubt that the PPP fell prey to the incorrect Soviet thesis of non-capitalist development, which I have discussed elsewhere.[14] They were not, however, singular in this, for at the time other theorists of the apparent radicalisation of the government appeared.[15]

The radicalisation of the state was inferred from the nationalisations referred to above, a literal reading of the government's self-advertised claim to be 'building socialism', and such foreign policies as recognising Cuba, support for Angola in its war of independence, strong anti-*apartheid* statements, support for the 'Arab cause' and loud advocacy of the need for a New International Economic Order. When examined closely many of these radical positions are revealed as mere rhetorical disguises. Thus the government supported the anti-Marxist forces in Angola for many years and did so until the very last stages of the war. Recognition of Cuba was a Caribbean-wide policy embracing the regimes of all the English-speaking territories on regional and nationalist premises. More important, however, this view of radicalisation ignored the question of internal democracy in Guyana and the present historical stage of the country's internal class struggles. From this standpoint, nationalisation can be seen as simultaneously aiding the consolidation of petty-bourgeois control of the state machinery and increasing the resources of the state as well as the leverage of the ruling party over the economy, and hence over the daily lives and livelihood of the masses. In other words these developments have made the state an improved instrument for the consolidation of ruling-class dominance.[16]

Co-operative socialism in crisis: the latest phase

Historically, the course of capitalist accumulation has always been marked by interruptions and crises as profit rates decline, disproportions in the structure of production emerge in an acute form, and generalised over-production (along with its inverse, under-consumption) characterises market relations. State ownership of the country's productive assets has not altered this relationship in Guyana, as the opportunity for profitable sale in the world market is still the fundamental determinant of the levels and patterns of accumulation. Since the mid-1970s a general crisis in the world economy has prevailed. The global dimensions of this situation (depressed commodity markets, stagflation, the energy crisis, a decline in private direct capital flows to the periphery, increasing Third World indebtedness, etc.,) are too well known to need discussion here. Of concern, however, is its impact on the economy of Guyana, and internal mis-management of this impact has added to the on-going crises of underdevelopment and dependency, creating in Guyana a production crisis of enormous proportions. The output data for rice, sugar, bauxite-alumina since the mid 1970s, as shown in Table 3.2, reveal the depth of this crisis.

As the world crisis began to make itself felt after the boom in sugar prices during 1974–75, the state's initial response was to opt for what is termed the 'easy' solution of extending its credit while hoping that the upheavals in world commodity markets would be temporary. Table 3.3 reveals the rapid expansion of the money supply over this period, the enormous growth of the public debt, and the dominant role the state has come to play in the utilisation of financial savings. As it has turned out this was, however, no ordinary downswing of the post-war business cycle, and more drastic measures had to be sought. They involved a mixture of severe deflationary policies, a frantic search for overseas balance of payments support funds (as well as the more normal project aid funds to stimulate expansion of the country's productive assets) and, wherever possible, the continued expansion of state indebtedness and the money supply as means of appropriating resources for the state sector. Each of these policies will be looked at briefly in turn.

The deflationary policies began in earnest in 1977 when the budget for that year proposed a cut in public expenditure of 30%, as well as price increases and the removal of subsidies on a wide range of working-class consumption items (many of which had been in existence since the colonial period).[17] A number of indirect taxes were also increased, e.g. customs duties, licences, fees, etc. This was later accompanied by a wage freeze in 1979, after the unilateral cancellation by the government of a previously negotiated three-year Minimum Wage Agreement with the national Trade Union Council for the period 1977–79.[18] Wage increases for the period after 1979 have been substantially below the official cost-of-living index, even though the index has been widely criticised as a severe underestimate of the level of retail prices and denounced as a 'political-cost-of-living index'.[19]

External balance of payments support funds were sought everywhere: from the OPEC countries, the Soviet bloc, Trinidad and Tobago and the IMF/World Bank. The latter two have turned out to be the major sources of aid. In August 1978 a one-year IMF stand-by facility of 15 million SDRs was negotiated, but the expected balance of payments turnround did not materialise, and in August 1979 recourse was had to a negotiated three-year extended fund facility. The government was, however, unable to meet the targets set under this arrangement and the facility was interrupted. The failure was due

Table 3.2 Selected economic indicators (physical)

		Output ('000 tons)					Electricity output (Mn kWh)			Vehicle registration		
Year	Pop. ('000)	Sugar	Rice	Dried bauxite	Calcined bauxite	Alumina	Total[a]	Residential	Industrial and commercial	Lorries	Tractors	Private cars	Other
1970	716	311	142	2,290	693	312	322	52	232	225	439	1,702	484
1971	729	369	120	2,108	700	305	329	59	240	120	242	1,454	357
1972	740	315	94	1,652	690	257	339	61	235	133	240	1,615	320
1973	750	266	110	1,665	637	234	362	70	252	185	375	1,413	503
1974	757	341	153	1,383	726	311	369	71	252	221	400	1,354	573
1975	766	300	175	1,350	778	294	383	77	252	455	693	1,377	911
1976	775	333	110	969	729	265	392	82	261	641	504	1,635	906
1977	783	242	212	1,001	709	273	336	79	215	246	289	626	508
1978	786	325	182	1,100	590	276	409	81	269	121	134	385	360
1979	790	298	142	1,117	589	171	407	95	245	246	239	626	508
1980	793	270	166	1,105	598	215	404[b]	98[b]	250[b]	63	400	643	480
1981	796	301	163	637	513	170	—	—	—	125	302	944	786

Notes.
(a) Figures do not total because of losses in transmission.
(b) Estimates.
Source. Ministry of Economic Development, Statistical Bureau, and Bank of Guyana Annual Reports.

Table 3.3 Selected economic indicators (monetary)

Period	International reserves (net) G$m	Investment % of gross domestic expenditure at market prices — Private	Investment — Public	Balance of payments G$m — Current account	Balance on monetary movements[a]	National debt G$ million — Total	External	Internal	Money supply $m — Total	% Change on previous year	Price indexes — Consumer prices	Terms of trade[b]
1970	54·6	12	10	−46·6	+2·1	267·2	160·0	107·2	165·0	7	100·0	104·5
1971	67·3	7	12	−13·2	+12·7	436·2	291·6	144·6	192·5	17	101·7	100·0
1972	89·7	8	12	−22·6	−22·4	499·6	316·2	189·4	231·7	20	106·7	96·4
1973	41·8	6	15	−123·4	+47·9	637·6	348·6	289·0	274·1	18	117·2	92·2
1974	105·4	7	17	−17·0	−63·6	672·7	403·6	269·1	317·9	16	140·3	110·9
1975	197·7	6	21	−35·2	−92·3	932·6	533·4	399·2	449·4	41	148·7	131·1
1976	−29·2	5	25	−350·8	+231·3	1,330·4	662·5	657·9	491·5	9	161·7	100·0
1977	−99·8	5	18	−251·1	+74·4	1,526·9	689·8	837·1	603·8	23	179·1	98·4
1978	−50·6	4	15	−72·3	−24·9	1,744·1	744·0	1,000·1	664·2	10	214·0	84·9
1979	−181·6	5	18	−208·1	+123·5	2,082·9	811·7	1,271·2	714·4	8	247·4	81·0
1980	−396·4	4	19	−300·4	+214·8	2,548·3	911·3	1,637·0	850·4	19	264·0	84·0
1981	−482·7	4	22	−558·0	+86·0	3,040·6	1,261·4	1,779·2	997·1	17	322·7	—

Period	Current government expenditure % by category[c] — Personal emoluments	Payments on public debt	Education	Health	Defence law and order[d]	Exports of goods and services as % of GDP at market prices	Imports of goods and services as % of GDP at market prices
1970	40	14	15	8	14	57·0	57·4
1971	43	14	15	8	13	57·8	54·7
1972	43	17	16	8	12	58·1	59·4
1973	40	21	15	8	12	52·2	68·9
1974	35	23	15	7	13	69·4	66·5
1975	33	21	14	7	14	80·8	80·2
1976	28	26	13	5	18	66·3	90·8
1977	35	30	17	5	15	62·5	78·8
1978	35	35	16	6	15	63·7	63·7
1979	30	39	15	6	13	60·1	70·7
1980	25	44	12	6	12	68·2	83·5
1981	—	—	—	—	—	51·1	79·2

Notes
(a) + = decrease in reserves; − = increase in reserves.
(b) Between 1976 and 1980 import prices rose from 100 to 192·4, while the export price index for calcined bauxite rose to 197·3, dried bauxite, 154·4, alumina 207·2, sugar 145·7 and rice 104·0.
(c) Totals in rows are not cumulative, since expenditure on personal emoluments duplicates expenditure on health, education, etc.

principally to the persistent inability of the state to achieve the required physical output targets in the three major export commodities, now completely under state control. A mixture of mismanagement, poor industrial relations and adverse weather, in descending order of importance, was chiefly responsible. In July 1980 a new three-year extended fund facility came into effect, involving 100 million SDRs from the IMF, in addition to a World Bank structural adjustment loan of US$23·5 million. The former was later increased to 150 million SDRs, or 400% of the Guyana quota of 37·5 million SDRs. By the end of 1981 the targets were again not being met and a 'withdrawal' from the new arrangements was pending. The targets/policies of the IMF/World Bank included the usual package: reduction in subsidies; elimination of price controls; increased interest rates; cuts in imports; elimination of commercial arrears in external trade; adjustment of energy prices in order to reflect their true costs; rationalisation of pricing policies in the state sector; cuts in social services; the establishment of commercial profitability as the main performance criterion for the state productive sector; the attraction of skilled personnel by way of appropriate salary payments; clarification of the status of the private sector through the official enactment of an Investment Code; and last, but by no means least, devaluation of the currency to its 'true' level.[20]

Two separate questions have been raised in response to these developments. The first is, how serious can the construction of socialism be if it is being built on the basis of World Bank–IMF credit? And the other is, why do the World Bank and the IMF support co-operative socialism in Guyana to the extent that they have and continue to? The second issue largely embraces the first, for the record shows that the IMF and World Bank have indeed been very generous with Guyana as compared with other governments of the region, e.g. Jamaica. Two explanations account for this. One is that the US government's influence over the IMF was designed to destabilise Jamaica under Manley while Guyana under Burnham was still regarded as imperialism's best bet. In addition, even before resort was made to the IMF in the extensive way it was after 1978, the government had already introduced policy measures (beginning with the 1977 budget) which clearly indicated that its method of dealing with the overall crisis was to give priority to correcting the balance of payments disequilibrium, and to achieve this by striving for a reduction in the levels of real consumption of the population in the expectation that, because of the high propensity to import, this would reduce imports and hence the level of foreign exchange expenditures.[21] Increases in the volume of exports, while hoped for, were not really expected to materialise rapidly enough in the context of the gross mismanagement of the state sector which by that time had become very noticeable. Such an approach coincided with IMF/World Bank thinking, and it is this which explains why, despite the socialist claims and the political realities of repression, rigged elections and political assassination of opponents of the government, international support at this level was so readily forthcoming.

Despite IMF/World Bank strictures on increasing indebtedness and easy money policies, the monetary data set out earlier reveal that they continued at a high level, although to a lesser extent than previously. While these international bodies continued to resist an easy money policy, it was the failure to reach the physical targets that created the biggest difficulties for the government in its relations with them.

The post-1977 deflationary policies did not move the economy out of its deep depression. The production data presented earlier reveal that while the country produced 369,000 tons of sugar in 1971, in 1980 it managed to produce only 270,000 tons and in 1981 only 300,000 tons, even though the current rated capacity of the industry is 450,000 tons. Dried bauxite, calcined bauxite and alumina output in 1980 was 1,600,000 tons, 602,000 tons and 211,000 tons respectively, though as far back as 1970 2,300,692 and 312,000 tons respectively had been achieved. Rice output in 1980 was only 160,000 tons, while the rated capacity of the industry was 250,000 tons. Accompanying these declines in real output was a virtual collapse of the various public utilities: electricity, public transport, sanitation, water supply, postal services and telecommunications. The social services collapsed under the strain, and the flight of people from the country has become a major feature of social life. The restrictions on imports have created grave shortages of food, spare parts and intermediate goods. The first two shortages have led to a serious depletion of the capital stock and a deterioration of services; the last, to a further curtailment of production.

Not unexpectedly, the production crisis, combined with the decline in the social services and widespread shortages of imported commodities, has meant a general reduction in the standard of living. In estimates used by some trade unions in Guyana the fall in real wages since 1976 has been of the order of 44%.[22] Industrial and other forms of social unrest have inevitably followed. The state's resort to repression in order

Guyana: 'co-operative socialism'

to contain the unrest has helped to reveal its essentially anti-working-class and unpopular character. Later in 1979 a near-insurrectionary situation developed, with many popular demonstrations against the regime. These were associated with the emergence of a new political party, the Working People's Alliance, one of whose leading activists, Walter Rodney, was assassinated by agents of the state in 1980.

This party has campaigned vigorously against the destruction of the economy, the abuse of human rights, the rapid growth of repression and the consolidation of a constitutional dictatorship. It has called for an end to racial separation, the creation of a Government of National Unity and Reconstruction comprising all democratic forces committed to the struggle to bring the dictatorship to an end, and the restoration of multiracial political democracy. These are advocated as the most immediate problems confronting the masses, as it is claimed that without democracy and multiracial unity no advance would be possible. These developments indicate an important consideration. The 'independence settlement' of 1966 did not constitute a transfer of power to the masses. The petty-bourgeois elements who acquired state power at that time did so as a result of a well documented Anglo-US manoeuvre against the then popular forces.[23] Their subsequent hold on state power has been based on force, as can be seen in the trail of rigged elections and unconstitutional practices ever since. In other words, a dictatorship has emerged as the crisis of production has become generalised into a social and political crisis.

The process of 'fascistisation' of the state which has paralleled the development of the general crisis in Guyana is best symbolised in the inauguration of a new constitution after the postponement of elections constitutionally due in 1978. At the end of 1980 'elections' were held under the new constitution, and the report of an international observer team revealed the crude and extensive rigging which went on.[24] Behind the change in constitutional status, however, the fascistisation of the state is revealed in the dramatic growth of three major dimensions of state activity: an increasing militarisation, bureaucratisation and ideologisation of state functions.[25] Danns estimated that in 1977 one in every thirty-five Guyanese was a member of one or other branch of the state security services.[26] Bureaucratisation has accompanied state penetration into the economy and the increasing need of the ruling political elite to regulate on its behalf all life in 'civil society'.

The ideologisation of state functions is seen in the 'nationalisation of all communication media' and in the almost literal prohibition of any independent, let alone opposition, voice. Control of the media has been used principally to promote an image of legitimacy for the government. In the face of falling living standards and increasing repression these efforts have not been particularly successful, certainly not within the country. As conditions have worsened, the media have been used more and more to 'attribute' responsibility for the economic crisis. Such attributions have been made to high oil prices, destabilisers, arsonists, Acts of God, and economic aggression by Venezuela (which has a territorial claim against the country). It is in this atmosphere that the pretexts for repression have been prepared.

Conclusion

Space does not permit detailed elaboration, but certain significant conclusions emerge from a study of the political economy of co-operative socialism in Guyana.

1. Co-operative socialism is an ideological rationalisation for the development of state capitalism in Guyana and for the creation of a new class of indigenous capitalists, 'fathered' in the first instance by the state. This is a general phenomenon in many Third World countries, reflecting the manner of their specific insertion into the system of international capitalism and the need for their present ruling elements to use the state in order to create sufficient national space for the growth of an indigenous capitalist class.

2. The particular form of the ideological rationalisation has been prompted by the broad expectancy of the masses, who have not only been schooled in militant anti-colonialism, but many of whom have long operated within Marxist-socialist political formations. This goes as far back as the 1953 election referred to above, when the first freely elected Marxist government in the British Empire came to office.

3. The importance of the state at this juncture is not to be taken as being principally premised on its position in the international capitalist structure referred to in (1) above. The state is first located in the internal class structure. This structure is, in the particular circumstances, extremely complex and variegated because of the absence of the fully developed class formations characteristic of our historical age (the working class and the bourgeoisie). The absence of a clearly hegemonic class has produced a certain 'fluidity' of political power, and the role of the state at this stage of the class struggle has become particularly vital. State power in the circumstances of Guyana does not simply indicate

the location of power in one or other of the major class formations of our age, but is itself an instrument of class formation.

4. In light of the above, co-operative socialism merely seeks a nationalist accommodation in the world capitalist order, and in no way is it premised on a fundamental alteration of the capital–labour relation, either internally or internationally.

5. Because the new economic class is based on a monopoly of political power, the traditional capitalist sequence of economic power leading to political power is reversed. Here political power is the principal instrument for the creation of the economic wealth of the emergent bourgeoisie. This factor is vital for understanding the ideological and structural functions of the state in Guyana.

6. Having come to power in the absence of free and fair elections or as the result of a popular social revolution, the present ruling clique lacks legitimacy. As the economic crisis has deepened it has sought to overcome this weakness through the fascistisation of the state. This process is therefore the ruling class's solution to the present internal crisis and the worldwide crisis of capitalism. From this standpoint it is not a simple, transient phenomenon, with all prospects of returning to a Westminster parliamentary type of democracy, which some treat as characteristic of the region.

7. The entire process described here does not, and indeed cannot, overcome dependence, although it alters certain characteristics of it. The principal alteration is, of course, the role of the state in the national economy. This has led to a challenge to the classic colonial division of labour and a fight for its replacement by a 'new' international division of labour which responds to the claim of the national ruling class in Guyana, even as that class needs to keep unchanged its internal privileges and system of domination. However, if reference is made to the criteria for evaluation set out earlier, we can conclude that dependency relations remain essentially the same. The same dynamic of producing what is not consumed and consuming what is not produced operates in the material sphere despite the growth of state property. Profitable sale in the world market is still the main determinant of accumulation. The capital–labour relation remains unchanged. No new science or technology is being created. The status and the right to work of the workforce has not been changed in any significant way. At the same time anti-popular, anti-participatory, anti-democratic and dictatorial conditions of social life continue to prevail, albeit clothed in the new rhetoric of 'co-operative socialism'.

Notes

1. One of these was operated by the British firm Booker McConnell Ltd, which accounted for about 90% of the crop; and at the time of its nationalisation in 1975 the other firm was a subsidiary of Jessel Securities Ltd.
2. Agricultural cultivation along with much of the settlement in the country occurs on a thin coastal strip which is below sea level and is subject to heavy tropical rainstorms. As a result costly infrastructural investment in drainage and irrigation is a prerequisite for agricultural cultivation. For every square mile of sugar cane cultivation, fifty miles of waterways are required.
3. See Clive Y. Thomas, *Plantations, Peasants and State: A Study of the Alternative Modes of Sugar Production in Guyana*, Institute of Social and Economic Research, Mona, Jamaica, 1982.
4. For a fuller discussion see Clive Y. Thomas, *Dependence and Transformation: The Economics of the Transition to Socialism*, New York and London, 1974.
5. See *ibid*.
6. Some of the ideas raised in this section are more fully discussed in Clive Y. Thomas, 'From colony to state capitalism: alternative paths of development in the Caribbean', *Transition*, V, 1981.
7. W. Haque *et al.*, 'Towards a theory of rural development', *Development Dialogue*, 1977.
8. See also the literature on the *Another Development* project published in various issues of *Development Dialogue* since 1975.
9. See L. F. S. Burnham, 'A vision of the cooperative republic' in L. Searwar (ed.), *Cooperative Republic*, Georgetown, 1970, and 'Declaration of Sophia', Address by the Prime Minister to the Special Congress, Tenth Anniversary of the PNC in government, Georgetown, 1975.
10. See N. Girvan, 'The Guyana–Alcan conflict and the nationalization of Demba', *New World Quarterly*, V. 1972, pp. 38–49, and 'Why we need to nationalize bauxite and how' in N. Girvan and O. Jefferson (eds.), *Readings in the Political Economy of the Caribbean*, Kingston, 1974, pp. 217–40, for a discussion of the bauxite nationalisations, and Clive Y. Thomas, *Plantations, Peasants and State*, chapter VIII, for a discussion of the terms of the sugar industry nationalisation.
11. G. Standing, 'Socialism and basic needs in Guyana' in G. Standing and R. Szal (eds.), *Poverty and Basic Needs*, Geneva, 1979, p. 24.
12. *Ibid.*, p. 44.
13. *Ibid.*, p. 52.
14. See Clive Y. Thomas, 'The non-capitalist path as theory and practice of de-colonization and socialist transformation', *Journal of Latin American Perspectives*, XVII, 1977, pp. 10–28.
15. See J. Mandle, 'Continuity and change in Guyanese underdevelopment', *Monthly Review*, XXVIII, 4, 1976, pp. 57–60, and *The Post-colonial Mode of Production in Guyana*, University of Guyana and Temple University, mimeo, 1978; and P. Mars, 'Cooperative Socialism and

Marxist scientific theory', *Caribbean Issues*, IV, 2, 1978, pp. 71–106. Critics of these views are E. Kwayana, 'Pseudo-socialism', Paper presented to a seminar of the University of the West Indies, Trinidad, 1976, and Clive Y. Thomas, 'Bread and justice: the struggle for socialism in Guyana', *Monthly Review*, XXVIII, 1, 1976, pp. 23–35.

16 For further discussion see Clive Y. Thomas, 'State capitalism in Guyana: an assessment of Burnham's Cooperative Socialist republic' in F. Ambursley and R. Cohen, *Crisis in the Caribbean*, London, 1983, pp. 27–48.

17 See University of Guyana Staff Association, *The Working People and the Economic Crisis: Towards a Solution*, Georgetown, 1979.

18 After much agitation in trade union circles it was agreed that the minimum wage would increase from G$5·40, to G$8·40, G$11·00 and G$14·00 respectively over the years 1977–79. These wages became, however, the maximum wage changes permitted in the state sector. No wage increases were allowed in 1979. In 1980 a 5% increase was permitted and in 1981 increases ranging from 7% to 10% and 10% to 12% for unskilled and skilled worker categories were allowed.

19 See University of Guyana Staff Association, *The Working People*. In its 1980–81 Report the Inter-American Development Bank also remarked in reference to movements of the official index, 'This relatively moderate inflation rate (16·5%) *seems rather curious* in an environment characterized by a 15% increase in the import price index and considerable monetary expansion in the domestic economy coupled with rather slow growth in real output,' p. 271; my emphasis.

20 The currency was devalued by 18% against the US dollar in 1981.

21 See Thomas in Ambursley and Cohen, *Crisis in the Caribbean*. This article cites some examples of their policy measures. It also points out that between 1976 and 1980 there was a gross population increase of 80,000 persons but migration, both reported and unreported, accounted for 62,000 persons. The net increase was therefore only 18,000 persons.

22 These estimates do not accept the government's cost-of-living index as reliable, since it is based on a 1956 household budget survey which includes many imported items banned from entry into Guyana. The suspicion is that this index is deliberately used because it understates the real rate of inflation.

23 See Cheddi Jagan, *The West on Trial*, New York and London, 1966. In this the details of the CIA plot to oust Jagan and replace his government with one led by Forbes Burnham are fully documented. The role of the US unions, Churches and State Department in destabilising the regime is supported by quotations from the reputable press and memoirs published by officials in the US government at the time.

24 International Team of Observers at the Elections in Guyana, *Report*, published by the Parliamentary Human Rights Group, House of Commons, London, 1980, reprinted by the Guyana Human Rights Association.

25 For a fuller discussion see Clive Y. Thomas, *The Rise of the Authoritarian State in Peripheral Societies*, Monthly Review Press,

forthcoming. The process of fascistisation does not mean that the state in Guyana is fascist. I have used this term for want of a better one, since the work referred to above on the authoritarian state, which I believe is the correct description of this state, is not at the time of writing generally available to the reading public. In other words the readers should place emphasis on the *process* while noting the methods of rule one associates with fascist-type regimes.

26 G. K. Danns, 'Militarization and development: an experiment in nation-building', *Transition*, 1, 1978, pp. 23–44.

Tony Thorndike

4 Grenada: the New Jewel revolution

Political milestones are often claimed but rarely justified. One which thoroughly deserves the title was the near bloodless revolution of 13 March 1979 on the tiny 133 square-mile island of Grenada and its offshore wards of Carriacou and Petit Martinique. Not that it was a precursor of revolutionary change elsewhere in the Commonwealth Caribbean or in the region generally, since it was a specific response to the particular political and economic circumstances of Grenada at the time. Nor does its importance derive simply from the overthrow of a dictatorial, discredited and dishonest regime, that of Sir Eric Gairy, although its demise was an essential precondition of what was to follow. Rather, the revolution was to be a continuing phenomenon which acquired considerable, and increasing, geopolitical and ideological significance, both regionally and hemispherically. Crucial to its understanding is that, from the start, it represented a conscious policy of contesting rigorously the state of dependency which so fundamentally conditioned the development and direction of Grenada's political economy and the lives of its 110,000 inhabitants.

The dimensions of dependency

For the purposes of analysis, dependency is seen as having three dimensions: the structural, the directed and the psychological. Although all three manifestations are closely connected, a concrete reality which the post-*coup* People's Revolutionary Government (PRG) fully appreciated, each represents a different set of problems demanding separate solutions.

The first dimension, the structural, is the principal dimension and, by its nature, is manifested both internally and externally. Since the basic structure of the economy was substantially determined by external market pressures over which the producers and society at large had little control, reform through planning and agro-industrial development within the physical constraints of Grenada's small size and population was the PRG's top priority. Concurrent with this was action to alleviate the ravaging effects of this structural dependency on the people. Their poverty, it was argued, derived as much from exploitation by a small Gairy-led local merchant- and banker-dominated class, dependent upon North American and Western European interests, as from externally directed dependency itself. Curbing the activities of this group went hand in hand with job opportunities programmes and greatly expanded social services provision.

Since the development and evolution of structural dependency is ultimately directed by external forces, then directed dependency, or the conscious actions of core or dominant economies to ensure the continuance of their privileged position both economically and politically, had to be equally challenged. To this end, the PRG, under Maurice Bishop's leadership, rapidly came to the forefront of those articulating Third World demands for a 'New International Economic Order'. Part and parcel of this was an insistence on its sovereign right to diversify trading patterns and sources of aid and imports, together with the right for its sovereignty to be respected by others as it would theirs. In Bishop's words,

> Small as we are, and poor as we are, as a people and as a country we insist on the fundamental principles of legal equality, mutual respect for sovereignty, non-interference in our internal affairs and the right to build our own process free from outside interference, free from bullying, free from the use or threat of force.[1]

Six principles of foreign policy were enunciated: the designation of the Caribbean Sea as a zone of peace; the right of self-determination recognised for all peoples in the region, particularly in the non-independent territories; acceptance of the principle of ideological pluralism, and an end to propaganda and 'economic and violent destabilisation'; an end to the arming and financing of counter-revolutionaries and 'anti-progressive' regimes; respect for the sovereignty, legal equality and territorial integrity of all countries in the region; and the freedom to join whatever international organisations or create any regional or sub-regional groups which were in the best interests of the people.[2] It was necessary to identify these principles because of imperialism.

> We contend, comrades, that the real problem is not the question of smallness per se, but [that] of imperialism. The real problem that small countries like ours face is that on a day-by-day basis we come up against an international system that is organised and geared towards ensuring the continuing exploitation, domination and rape of our economies, our countries and our peoples. That, to us, is the fundamental problem.[3]

Thirdly, as Grenadians had passively accepted an inferior position relative to the more powerful, and had adopted their values and preferences since there seemed no alternative, psychological dependency had to be undermined. In its place was planned a new sense of Grenadian identity, patriotism and pride. Each Grenadian would be expected to identify individually with the revolution and its aims and to be an integral part of the decision-making process. It was argued that constitutional patterns inherited from colonialism and developed in quite different political and historical circumstances not only conditioned the population to accept dependency but also helped disguise it by a cloak of constitutional 'legitimacy'. In other words, there was a form of 'false consciousness': the people believed the Westminster model to be the only way political activity and thought should be conducted and consequently distrusted any constitutional framework designed by themselves for themselves to suit their particular country. Viewed from this perspective, therefore, an imposed constitution and legal framework were seen as designed to perpetuate 'colonial' attitudes of subservience. Hence the parliamentary system bequeathed by Britain at independence in February 1974 was abolished. Instead there is experimentation with a radically different political framework labelled 'participatory democracy', the institutionalisation of which is still evolving.

> Today in Grenada, Parliament has moved out of town and into the communities. Government has escaped ... and spread into community centres, school buildings, street corners, market places, factories, farms and workplaces around the country. Political power has been taken out of the hands of a few privileged people and turned over to thousands of men, women and youth ... in every nook and cranny of Grenada, Carriacou and Petit Martinique.[4]

While it is too early to assess the effect of the education and mass mobilisation programmes which have been initiated alongside this new framework, and which aim to develop through example and persuasion a new and more 'progressive' set of social and political values, there is no doubt that the general philosophy of the PRG and its ideological and policy preferences command widespread support, particularly among the peasantry and working class.

The Grenadian revolution, therefore, goes far beyond the overthrow of tyranny. Not unexpectedly, its external impact attracted the world's attention and accorded it a far greater importance than Grenada's small size would suggest. Although the PRG's foreign policy is labelled that of non-alignment, a suspicious region and USA became first apprehensive, then angry, when it turned to Cuba for military aid to help the 2,000-strong People's Revolutionary Army (PRA) and the larger Militia secure the revolution. This was followed by substantial Cuban economic assistance, most notably for the Point Saline International Airport. In an area of already heightened sensitivity, graphically characterised by successive US State Department spokesmen as a 'circle of crisis' and 'a sea of splashing dominoes',[5] the US feared the airport would be used for Cuban and Soviet military purposes. Strong US disapproval was met by stubborn insistence on the PRG's right to determine the nature and direction of Grenada's international relations as it wished. Grenada's neighbours in the Organisation of East Caribbean States (OECS) shared Washington's concern, particularly in the immediate aftermath of the revolution. They feared widespread revolutionary instability, with other radical groups attempting to overthrow governments and then to rule without elections in the name of 'the people'.[6] However, as subsequent events made clear, not only were the left-wing activists in the region nowhere near as strong as their loud praise for the PRG suggested, but all subsequent elections in the Commonwealth Caribbean, such as in Jamaica, Trinidad and Tobago, Dominica, St Kitts-Nevis and St Vincent, confirmed the continuing popular commitment to ingrained and traditional political institutions and, furthermore, showed a trend towards the centre right. Grenada, in this context, was a singular and highly specific event.

The seeds of revolution

Any analysis of why the revolution was unique in the Commonwealth Caribbean must take account of Gairyism and the radical opposition which it spawned. Noted for his charm and sartorial elegance, Gairy's background was that of a primary schoolteacher who in the 1940s had emigrated to Aruba to work in the oil refinery installations. Deported in 1948 for union activities, he returned home and formed the Grenada

Manual and Mental Workers' Union and the associated Grenada United Labour Party (GULP) the following year. His political fortunes were established by the leadership of a highly successful strike followed by electoral victory over the political representatives of the Grenadian middle classes in elections fought in 1951 for the first time on universal suffrage.[7] The working-class challenge, which he represented, was not only based on increased class consciousness but also sprang from their tradition of independence from, and lack of subservience to, the bourgeoisie. Labelled 'agroproletarians', i.e. signifying a blend of peasant landholding and casual plantation work, 41% of all farmers (the largest occupational group) owned and worked one acre or less, and 46% between one and five acres.[8] To such an electorate, Gairy's flamboyancy and open flouting of the pro-planter colonial law was irresistible. His satirical treatment of the elite after they had rebuffed his efforts to gain social recognition and be treated as an equal, and his mysticism, were added ingredients. The latter was taken seriously: for instance, in late 1952 Gairy called on God to send a sign that he favoured a strike.

> That night there was a downpour, heavy even by Grenadian standards. The road between St Georges and Gonyave was blocked by fallen rock, which many regarded as a sign of divine support ... The Public Works Department tackled this roadblock with unusual energy, but took a fortnight to remove it. With Gairy's divine sign, and a wave of awe sweeping Grenada, police took up protective positions ...[9]

His hold on them was not so much through GULP but rather the use of the 'crowd'. His ability to mesmerise and channel the crowd's emotions for short but intensely emotional periods of time caused one political scientist to suggest the concept of a 'hero' to describe Gairy's role.[10] By contrast, the Grenada National Party (GNP), formed in 1953 and led by a lacklustre barrister, Herbert Blaize, represented the urban middle class and the merchant interest. He and his party were able to gain political influence only with the aid of external pressures, as when, in 1961, Gairy was removed from office by the British for 'squandermania' and electoral malpractice.[11] But Gairy resoundingly won the 1967 elections and remained in power for twelve more years. However, his dominance among the rural peasantry progressively weakened as fewer and fewer slices of a diminishing economic cake were allocated to them and economic mismanagement and corruption grew. Also, Gairy steadily distanced himself socially and politically from his power base, encouraged by many of the merchants and other sectors of the middle class who gave lip-service support because of his widespread powers of patronage.

Since the GNP was the only vehicle of opposition, it grew despite its ineffectual leadership. By the beginning of the 1970s it included for tactical electoral reasons a young, educated urban group who styled themselves the 'Committee of Concerned Citizens' (CCC). A month after the February 1972 elections, again won by Gairy, there was established JEWEL, 'Joint Endeavour for the Welfare, Education and Liberation of the People'. Led by an ex-GNP candidate, Unison Whiteman, it was based on a rural farming co-operative and sought, with some success, to mobilise the agro-proletariat.[12] In October 1972 the CCC joined the recently formed Movement for the Advancement of Community Effort, led by Maurice Bishop, Bernard Coard and Kenneth Radix, to form the 'Movement for Assemblies of the People' (MAP). Predominantly intellectual and professional in content, and based in the capital, St Georges, MAP was particularly interested in the abolition of the 'inappropriate' two-party system in favour of 'people's assemblies' modelled on the 'ujamaa' villages of Tanzania. In an interview Bishop said:

> We envisage a system which would have village assemblies and workers' assemblies. In other words, politics where you live and politics where you work ... elections in the sense of the elections we now know would be replaced by Assemblies at different levels. Grenada is small enough for this type of mass participation.[13]

In this scheme, a National Assembly made up of representatives from lower-level assemblies would elect a council to put its decisions into practice. Members of the council would be on committees formed to head government departments. This new system, it was argued, 'will end the deep divisions and victimisation of the people found under the party system'.[14]

Both JEWEL and MAP had similar broad aims and had been inspired by the development of the Black Power movement in the late 1960s and early 1970s, especially that in nearby Trinidad and Tobago. There were also elements of Rastafarian culture in their philosophy. They finally merged in March 1973 to form the New Jewel Movement, or NJM, which developed a more socialist trend in ideas and policies. Particular emphasis was put on the formulation of 'principled positions' on a wide range of issues which could not be compromised.[15] Also, whereas neither JEWEL nor MAP had as their objective the formation of political parties, the NJM decided to work within the existing political system until the 'Tweedledum and Tweedledee situation', with 'two parties which were two sides of the same coin',[16] could be replaced. Within a short time it succeeded in effectively supplanting the GNP as the main vehicle for anti-Gairy opposition. The issue was that of independence.

Whereas the GNP and its allies, the Chamber of Commerce and the Employers' Federation, argued that Grenada's economic position was too precarious to support independence, the NJM saw it as leading to higher standards of living, impossible to achieve under Gairy.

> Independence must mean better housing for our people, better clothing, better food, better health, better education, better roads and bus service, more jobs, higher wages, more recreation – in short a higher standard of living for workers and their children.[17]

Consequently, it saw Gairy's quest for independence as 'an insecure opportunist move designed to strengthen the grip of tyranny and corruption'.[18] By mid-1973 strikes were widespread, and petitions against independence – without a referendum – were sent to London. The economy suffered badly and Gairy brutally overreacted. His 'police aides', such as the feared Mongoose Men and the Night Ambush Squad, were openly encouraged, in his words, to 'cinderise' his opponents. Bishop's father was one of those killed, whilst Bishop himself, Coard, Whiteman and others were severely beaten in the cells of Grenville police station. All the opposition groups came together as the 'Committee of Twenty Two' but failed to prevent independence under Gairy.[19] Only strong regional pressure forced the post-independence publication of the Duffus Report, which catalogued the brutality of the police aides, openly joined by some policemen, but little response was made by the government in the way of reparation or reform.

After independence, the GNP's only remaining political base was Carriacou, Blaize's home. As the NJM was now in the forefront of opposition (and persecution), Gairy stressed its alleged Marxist element and justified further repression in the name of anti-communism. The 1976 elections, although fraudulent, saw the NJM as the senior partner in a People's Alliance which captured six of the fifteen seats. But Gairy's 'bewitchment' of the 'crowd' and, above all, his mystic powers and conviction that he was 'appointed by God to lead Grenada' maintained his confidence, despite the fact that internationally his reputation became steadily more discredited.[20] Indeed, his self-confidence was such that when he left for New York in mid-March 1979 to persuade the UN Secretary General, Kurt Waldheim, to create a UN agency responsible for UFOs and other cosmic phenomena, he gave orders for the NJM leadership to be assassinated. Pressure on the NJM had been building up for some months, with police raids and victimisation of supporters. This persecution not only forced the NJM leadership to consolidate its organisation but also led to the development of a clandestine wing. Trained in insurrectionary activity and equipped with arms smuggled in from the USA in oil drums, it was to form the nucleus of the post-revolutionary PRA.[21] Forewarned by some police allies, action was decided upon: within twelve hours what was described as a 'ragbag' army of only 200 ensured complete control.[22] As to why Gairy chose that moment to rid himself of effective opposition, given that since the 1976 elections Bishop had been the official Leader of the Opposition and a figure of national and regional significance, it may be surmised that he hoped that the new right-wing administration in Washington would not be so shocked at the removal of 'communists' as would normally be the case.[23]

Three days later a Provisional Revolutionary Government was formed, with nine NJM members, two GNP and three others. Led by Bishop, with Coard as Deputy Leader and Finance Minister, it was later expanded to twenty-three, the NJM comprising two-thirds, and given its present name. A Cabinet of seven was established, incorporating six of the eight members of the NJM Political Bureau.[24]

Faced with regional opposition from some quarters, US disapproval and a real threat of a mercenary-led counter-revolution by Gairy, organised from San Diego and Miami,[25] security was an immediate priority. In their search for help and recognition Bishop and his comrades-in-arms first sought military assistance from Grenada's traditional allies, Britain, Canada and the US.[26] All refused recognition of the regime, in contrast to Jamaica and Guyana, as well as military aid. Britain and Canada pledged non-intervention but the US remained ominously silent. Faced with what the PRG saw as a desperate situation, Cuba was approached. The result was critical: it set Grenada on to a foreign policy course which increasingly stressed the importance of the socialist bloc. Within a month, and following an emergency shipment of arms, ambassadors were exchanged, the Cuban ambassador being the only resident foreign diplomat in Grenada. Wide-ranging aid programmes were negotiated, of which the airport project was to take the lion's share, but medical aid was to be the most immediately beneficial.[27]

Revolution: by the people, for the people

Once secure, the PRG was able to mount its struggle against dependency in its three manifestations and embark on economic reconstruction. Structurally, it was decided to retain a mixed economy, despite the government's socialist preferences. No properties or businesses were nationalised except the electricity corporation and telephone company, and hotels and some land once owned by Gairy and his Attorney General, Derek Knight. Bishop also insisted that the growing number of state enterprises and co-operatives would have to 'compete with our private sector for profits' in an effort to avoid inefficient government monopolies which plague many socialist countries.[28] But the emphasis on central planning, more import controls and the dominance of the state sector 'which will lead the development process' helped to lead to the relative and, by 1981, the absolute decline of the private sector. To counter this, Coard announced preferential tax rates to companies, whether Grenadian or from elsewhere, prepared to invest in Grenada. Other incentives to private investment, announced in 1979, were reaffirmed.[29]

But none of the three sectors, private, state or co-operative, was to be immune from state mobilisation campaigns. Following the Cuban model, the government, as a 'workers' government which has exactly the same wishes and goals as all militant and progressive trade unionists',[30] began to press for Production Committees, composed of management and union representatives, to be established in every workplace. Responsible for drafting, discussing and putting into action work plans, and for monitoring management to prevent 'abuse of power', they would be paralleled by Disciplinary, Education and Emulation Committees. The latter would set production targets, 'devise and organise brotherly and sisterly competition' and reward the efforts of exemplary workers.[31]

The underlying philosophy was that, internally, structural dependency could be tackled only if the people were led and mobilised in the name and cause of their common good. Mass organisations, such as the National Women's Organisation, National Youth Organisation and trade unions supplemented the PRG's mobilisation and education efforts, particularly through the mass literacy campaign organised by the Centre for Popular Education. Price controls of essential commodities and a widespread development of social services boosted living standards and encouraged the acceptance by the people of self-help schemes. By 1982 all enjoyed free medical and dental care, free secondary education and the opportunity to rebuild and repair houses with National House Repair Programme grants. Despite massive inherited debts, help was given to establish labour-intensive enterprises, especially agro-industries and state farms, and, above all, to make Agricultural Marketing Boards efficient and improve feeder roads. Idle land, largely owned by Grenadians overseas, was taken over on short-term lease arrangements and run by the Land Reform Commission, using unemployed labour. This concentration on agriculture was aimed both at eliminating the country's food import bill and at encouraging more young persons to join the existing 5,000 small farmers. Within three years over 3,000 jobs were created, training was organised for all sectors of the economy, schools were refurbished and rural co-operatives established. Such developments helped ensure popular enthusiasm for the revolution and maintain its momentum, despite successive hurricanes and periods of extensive flooding.

However, world price movements resulted in a fall of 13% in Grenadian export prices in 1980–82 while prices of its imports rose by 12%. This led to a very substantial foreign trade deficit despite the purchase by the increasingly important Marketing and National Importing Board of Cuban materials at below-cost prices. Compounded with a 26% fall in tourist receipts over the same period, variously ascribed to the world recession and the 'continuous unprincipled and blatant propaganda campaign' waged by the North American media against Grenada, the constraints induced were considerable.[32] It should also be noted that in the absence of a central bank and a national currency (the East Caribbean dollar being administered by a Currency Board in St Kitts, with a fixed parity to the US dollar) planning is a difficult exercise. It is creditable, therefore, that the rate of economic growth has averaged 2·4% per annum in the first three years of the revolution, compared to an average of *minus* 3·2% from 1970 to 1978.[33] Although greatly increased efficiency and an end to corruption played a part in this, an important factor was widespread construction projects, centring upon the International Airport. The ramifications of these were to form the core of the challenge to directed dependence.

The USA and the airport issue

The airport project dramatically illustrates the Cuban link. Cuba is now Grenada's closest friend and ally, and represents a crucial element in the PRG's challenge to those external forces it perceives to be the sources and purveyors of dependency. The intimacy of relations between the two states is evident at all levels and is inevitable given the present extent of Cuban support and assistance. Subsequent US-led opposition and discrimination against Grenada has served only to strengthen the links. Indeed, it is reasonable to suggest that it was clumsy US diplomacy, rather than any particular predisposition on the part of the PRG, which led to the rapid deterioration of its relations with Grenada. Perceiving socialist and neo-Marxist influences in the new government, and alert to the possibility that the PRG might approach Cuba, the US ambassador (resident in Barbados) despatched a typed warning to St Georges a week after the seizure of power. It contained the following 'advice':

> Although my government recognizes your concerns over allegations of a counter-coup, it also believes that it would not be in Grenada's best interests to seek assistance from a country such as Cuba to forestall such an attack. We would view with displeasure any tendency on behalf of Grenada to develop closer ties with Cuba.[34]

Bishop angrily retorted that Grenada was 'not in anybody's backyard and we are definitely not for sale'.[35] At a subsequent mass rally he went on to accuse the USA of propaganda 'destabilisation'.[36] Later, accusations were made that the CIA was implicated in a bomb attack at a PRG celebration rally in June 1980.[37] Aimed at the PRG leadership, it resulted in two deaths and several injuries among bystanders. Although the allegations were officially denied by the USA, its increasingly frequent public attacks on the PRG and the evidence of past CIA malpractice led many in the region and elsewhere to believe that there was at least a grain of truth in the allegation.[38]

It is clear that mutual suspicion between the USA and Grenada has bred antagonisms which, in turn, have increased suspicion. The constant and stubborn advocacy of the leadership's 'principled positions' over a variety of issues angered not only the USA but most of Grenada's neighbours. Extreme sensitivity to any criticism led to strongly worded attacks which exacerbated the tension. Barbados was a particular target because of its pro-US stand and demands for elections, and diplomatic relations were nearly broken. The invective used, coupled with revolutionary rhetoric and suppression of internal 'counter-revolutionary' criticism, served only to convince the USA and Britain further that a Marxist-Leninist state was in the making. As an alarmed British Foreign Office Minister remarked:

> Grenada is in the process of establishing a kind of society of which the British Government disapproves, irrespective of whether the people of Grenada want it or not.[39]

For its part, Grenada was elected to the Bureau of the Non-aligned Movement at the Havana conference in September 1979. In Havana, Bishop also associated Grenada strongly with the Sandinista victory in Nicaragua, later visited that country and welcomed one of the junta, Daniel Ortega, to the first anniversary rally in St Georges. Addressing the second Congress of the Cuban Communist Party in December 1980, Bishop stressed that:

> We give our solemn pledge that whenever circumstances require, we will unhesitatingly fulfil our internationalist responsibilities. Imperialism must know and understand that if they touch Cuba, they touch Grenada, and if they touch Nicaragua, they touch Grenada.[40]

Concurrent with this was a successful campaign in both the UN and OAS against President Carter's plan to establish a Caribbean military task force in response to charges that the Soviet Union had 'combat troops' in Cuba. The PRG also supported Puerto Rican independence and called for the return of the Guantanamo base to Cuba.[41]

Tension rapidly grew, with each party issuing none too veiled threats. Feeling increasingly vulnerable, the security-conscious PRG leadership became progressively paranoiac in its perceptions of US intentions. Every US action and statement was perceived to mean that the American colossus was bent on removing the PRG. Only a further emphasis upon Grenada's revolutionary credentials would do, the aim being to maintain, if not increase, the support of its socialist backers. Links with the Soviet Union, Libya and the German Democratic Republic were developed and aid programmes agreed. They were later joined by Iraq, Syria, Hungary and other non-traditional suppliers.[42] Also, on every international issue, beginning with Afghanistan, Grenada sided with the Soviet bloc, particularly in the UN. In the meantime, limited contact with Washington was being maintained. Letters were sent to President Reagan personally. They were not answered. A final attempt to maintain contact was made in a letter of August 1981. Again unanswered, it was published by the PRG, concluding with a warning that

> If you should allow this letter also to go unanswered... then we shall have to conclude that your government does not desire even normal or minimum relations... in which event we would be obliged to consider further measures necessary to consolidate and to defend the social, economic and political transformation process...[43]

US concern naturally centred upon Cuban influence – not surprising, given the PRG's empathy with the whole tenor of the Cuban revolution – and Cuban aid. Outstanding as this aid was in its scope, it is hard to see how the PRG would have survived as a revolutionary government without it. That explains why its friendship with Cuba is non-negotiable. At times, up to 370 Cubans have been engaged in construction, fisheries, health and agriculture. A new agreement was signed in September 1981 involving yet more substantial levels of aid in a variety of areas. As before, most was allocated for the airport project, on which Washington's fears became centred.

The US position was that the 9,800 ft runway was manifestly too large for the modest air traffic serving Grenada. In evidence to the Senate Foreign Relations Committee in December 1981 Assistant Secretary of State Thomas Enders asserted that 'all types of Soviet aircraft' could use it 'to land and refuel', particularly on the Cuba–Angola route. The existence of the airport, and its potential use by the Soviet bloc, were given as a reason for the Senate to ratify the sale of sophisticated F-16 fighter aircraft to Venezuela. It was pointed out that all that country (and much else besides) could be reached by Soviet Mig-23 aircraft operating from Grenada, a squadron having allegedly arrived in Cuba.[44] Grenada countered these allegations by saying that the airport had been planned since 1926 and had been the subject of several feasibility studies in the 1950s and 1960s. It was, it said, absolutely necessary for tourism and economic development. The existing airport was small and geographically remote, whereas its replacement was very near the major tourist area and convenient for St Georges. The runway was only slightly larger than Aruba's, a Dutch island half Grenada's size, and considerably smaller than those serving the Bahamas or Guadeloupe.

The project was estimated to cost US$71 million to construct. Cuba agreed to supply heavy equipment, steel, cement, fuel and up to 250 workers. Together valued at approximately US$40 million, the PRG was left to raise the remainder, mainly from the European Community and radical Arab states. US pressure on potential donors, particularly the European Community, was intense. The USA attempted to sabotage the co-financing conference hosted by the European Community with the aim of raising US$30 million for the project, and effectively prevented endorsement by the World Bank of Grenada's public investment programme, most of which was connected with the airport. But Grenada won the propaganda battle: as one black US lawyer remarked at a large pro-Grenada rally in New York:

> This US pressure on the airport project is perhaps the best thing the Americans have done for the PRG since the March 13 overthrow of Gairy.[45]

The US response was not only to refuse to answer letters. It virtually cut off diplomatic relations by excluding Grenada from the list of territories to which a new US ambassador in Barbados was accredited. It also refused the credentials of the PRG's youthful and radical ambassador in Washington, a twenty-eight-year-old woman. More seriously, threatening military manoeuvres were mounted. In October 1981 operation 'Ocean Venture 1981', a large naval exercise off the Puerto Rican coast, involved the invasion of 'Amber and the Amberines' (Grenada and the Grenadines) to depose its 'unfriendly' government and maintain an occupation until elections were called.[46] The admiral in command publicly referred to Cuba, Nicaragua and Grenada as 'practically one country and that the situation in respect of those countries is a political-military one'.[47] On the economic front the US insisted that a US$4 million aid package to be administered by the Caribbean Development Bank for the OECS states should exclude Grenada. The OECS angrily refused on principle to accept such discriminatory terms and the loan was not accepted, although Grenada's fellow members were careful not to associate this support with any approval of the regime's ideological posture or its Cuban links.[48]

Britain followed by enacting a similar ban on its post-hurricane aid package to WINBAN, the banana producers' organisation, which includes those in Grenada. Its justification, on the grounds that the PRG had not held elections, was particularly hypocritical in that at the same time British military aid was resumed to Chile, which had firmly set its back on such democratic practices. The ban was strongly resented by the small independent banana growers, who, like most of their fellow countrymen, still retained considerable affection for Britain and its royal family. Military aid in the form of personnel carriers was also subsequently refused; in both instances it was assumed that the

Grenada: the New Jewel revolution

Conservative government in London had been willing to bend under pressure of a new hard-line Republican administration. By contrast, France and Canada increased their aid, and were joined by Mexico.

The US, however, was unsuccessful in its bid to prevent multilateral aid from being granted by the IMF and the European Community. Coard denounced the US in the IMF as a 'political prostitute' and openly questioned in Brussels whether 'the EEC countries were going to allow their foreign policy to be dictated to by the United States'. In the event, although the US was able to insist that IMF stand-by credit and hurricane relief was not to be used for the controversial airport project, it was unable to prevent the European Commission from declaring the airport to be 'essential infrastructure'.[49]

The obvious tension revealed an emerging paradox. Whereas Grenadian foreign policy was aimed at securing the revolution to enable it to challenge dependency, in the process its security was becoming weaker. Following the electoral defeat of Michael Manley in Jamaica in October 1980, and strong protests to Guyana following the assassination of the radical activist Walter Rodney a few months earlier, it was virtually isolated in the Commonwealth Caribbean. In Latin America it could by 1982 count on only two friends, Mexico and Nicaragua, since Venezuela had lessened its earlier support. In the region it responded by withdrawal from multilateral initiatives, despite its support of both CARICOM and the OECS.

Democracy or people's power?

Another pressure against Grenada from the region, again shared by the US and Britain, was over elections. Predictably, the PRG's 'principled position' on this aspect of psychological dependency was absolute, dating back to the days of MAP. Indeed, one of the more outspoken and emotional members of the PRG, Selwyn Strachan, called the seizure of power 'the fairest election' Grenada had ever had, since it was achieved on the basis of 'one man, one gun'.[50] Nonetheless concern was expressed that, no matter the extent of a principled commitment to some form of democracy other than parliamentary representation, PRG support should be tested and minority views permitted to be expressed notwithstanding the evident popularity of the regime as witnessed in the enthusiastic endorsement of its policies at mass rallies. Barbados went so far as to offer help in the form of returning officers to prepare voters' lists and organise a ballot. The offer was rejected in strong terms by the PRG, which also moved increasingly to stifle internal opposition to its views on the grounds that it 'represented the CIA in their plot to overthrow the revolution'.[51] The methods it adopted were either direct government action (including house arrest or detention for an indefinite time, the surrender of passports and telephones, or the confiscation of assets such as cars) or 'unofficial' intimidation by NJM or other youth groups. The general curtailment of civil liberties and promulgation of 'People's Law' by the Cabinet was necessary, it is argued, because of the activities and designs of the imperialist enemy without and its counter-revolutionary allies within.

In any event, it was pointed out that human rights of another, higher kind, those of welfare provision and protection against exploitation, as in the wholesale removal of 'anti-worker' ordinances from the statute book, were of greater importance and relevance to the revolution and the struggle against dependency. When deep-rooted colonial values are being changed there can be no alternative 'undemocratic' voice. Hence newspapers not controlled by the government, such as *Torchlight* (1979) and *The Grenadian Voice* (1981), have been closed for 'slander' and 'imperialist propaganda'. The publishers of the latter, the so-called 'Gang of Twenty Six', were arrested for alleged implication in a CIA plot to overthrow the PRG. Their number included the highly respected anti-Gairy journalist Alistair Hughes and a former PRG Attorney General.[52] The PRG's acute sensitivity to internal criticism was, in part, caused by persistent reports throughout 1981 and 1982 of an impending invasion by Cuban exiles organised by a 'Patriotic Alliance' of anti-PRG Grenadians based in Trinidad. Nonetheless it led a pro-PRG regional journal to editorialise:

> For us the curtailment of civil liberties should not be neglected in any country of our Caribbean region because they are non-existent elsewhere. Therefore, in this season of intense, hawkish pressures on his government, Mr. Bishop needs to act with restraint as the PRG continues its programme of educating the Grenadian people about their real enemies. A vulnerable island state like Grenada must appreciate how a policy of unchecked political detention can be exploited by its opponents and used as an excuse for external aggression.[53]

Notwithstanding this criticism, the PRG pressed ahead with its attack on psychological dependency. Faced with close popular identification with inherited constitutional forms, not surprising when the experience of formal governmental institutions has been as long as

West Indian colonisation itself, Bishop and his colleagues repeatedly denounced the 'two-seconds democracy' of the 'discredited' Westminster model. They argued that

> the type of democracy where people walk into a ballot box and vote for two seconds every five years is not real democracy at all.[54]

There will never, therefore, be any elections of the traditional type. Instead, parish councils became the focus of popular discussion where the people could discuss policy, question the leadership, civil servants and heads of government enterprises and let their views be generally known on both broad national issues as well as specific problems in the parish. Originally for NJM members only, this process was a continuation of pre-revolution NJM policy and, as the revolution consolidated, the councils were opened to all Grenadians who stood for NJM ideals. They became so popular and heavily attended that by late 1980 there was further dilution with the creation of zonal councils, the zone being a subdivision of the parish. In addition, the mass organisations have their own parish-level councils. The parish council, zonal councils and all the mass organisations within a parish are serviced and co-ordinated by a Parish Co-ordinating Body. This in turn interacts with the local NJM party organisation and, when necessary, with the national party. In turn, the national party is led and co-ordinated by the Political Bureau, most of whose members are in the Cabinet.

Clearly, this structure of 'People's Power' is democratic centralist, where views and opinions are transmitted between all levels.[55] It ensures the active involvement of thousands of people in the decision-making process, as, for instance, discussions on the 1982 budget showed.[56] As the *Free West Indian* newspaper put it, such 'ongoing participation by the people' involves a 'deep grasp' of what 'imperialism is in concrete terms', and prepares them 'for the broader objectives of building socialism' and an 'economic and social transformation'.[57]

Conclusion

In conclusion, it is probable that Grenada will remain unique in the Commonwealth Caribbean for the foreseeable future. However, this uniqueness, coupled with the commitment, at least rhetorically, to the cause of revolution in the Caribbean and Central America, makes it appear vulnerable. Traditional sources of aid and friendship have been cut off; economic gains have been made by means of effective popular mobilisation and planned management, but with temporary economic sacrifices due to the lack of credit from some multilateral lending agencies. But although Grenada is distant from its new-found friends, it is highly likely that military action against it would invite their intervention. This would most seriously affect the promotion of a Zone of Peace, so earnestly sought by the small and vulnerable Eastern Caribbean states.

To the PRG this is a risk that has to be accepted. After all, as one prominent Marxist PRG sympathiser warned:

> Sovereignty and national fulfilment can only be achieved as a result of a triumph over imperialism. To achieve this triumph the masses of the people must be made fully aware of and understand the issues involved, and inspired to the necessity of waging a struggle against imperialism, whether or not this involves the endurance of hardship and a temporary decline in their standard of living. The anti-imperialist struggle cannot be fought by the people in any country in isolation from the world struggle against imperialism, for the pursuance of which international alliances, of both a political and economic nature and, if necessary, a military nature, must be forged.[58]

It is therefore reasonable to assume that continued ostracism of Grenada due to its uncompromising pursuit of its policies and its verbal attacks on its critics will, in the long term, serve only to obscure the main purpose of its revolution. It was, and remains, the struggle against dependency in its three dimensions: the structural, the externally directed and the psychological. Through this, its aims of working for a mode of genuine self-government, social justice and a united society, breaking out of the mould of a colonial past, are being slowly realised. There have been mistakes, and the constant use of the CIA as a scapegoat cannot always be repeated, or believed, in the future, whatever the provocations. Nonetheless the Caribbean and Central American region may find that it has much to learn from the Grenadian revolution, whatever the political labels ascribed to it.

Notes

1 *Forward Ever! Against Imperialism and Towards Genuine National Independence and People's Power*, Address by Comrade Maurice Bishop, Prime Minister, People's Revolutionary Government of Grenada on the occasion of the First Anniversary of the Grenada Revolution, 13 March 1980, Havana, 1980, pp. 16–17.
2 *Ibid.*, pp. 19–23.

3 *Imperialism is the Real Problem*, Address to the Conference on the Development Problems of Small Island States by Comrade Maurice Bishop, 13 July 1981, St Georges, mimeo., p. 1.
4 *Is Freedom We Making: The New Democracy in Grenada*, St Georges, 1982, p. 22.
5 *U.S. News and World Report*, 5 September 1981.
6 H. S. Gill, *The Grenada Revolution: Foreign Policy and Caribbean Geopolitics*, Paper delivered to the Conference of the LDCs of the Caribbean: Domestic, Regional and International Perspectives, Antigua, June 1980, mimeo., p. 16. St Lucia's Premier, John Compton, went so far as to suggest privately that British troops should be called upon to reverse the insurrection.
7 P. Emmanuel, *Crown Colony Politics in Grenada, 1917–1951*, Barbados, 1978, pp. 178–85.
8 M. G. Smith, *Stratification in Grenada*, Berkeley, 1965, p. 14.
9 M. G. Smith, *Plural Society in the British West Indies*, Kingston, 1965, p. 290.
10 A. W. Singham, *The Hero and the Crowd in a Colonial Polity*, New Haven, 1968.
11 Specifically, Gairy was dismissed after he was strongly criticised by a Commission of Enquiry for contravening expenditure regulations and forcing civil servants by the threat of dismissal 'to commit or condone improprieties or irregularities in public expenditure'. *Report of the Commission of Enquiry into the Control of Public Expenditure in Grenada During 1961 and Subsequently*, Cmnd 1735/1962, London, HMSO.
12 *Trinidad Express*, 10 March 1974.
13 W. R. and I. Jacobs, *Grenada: The Route to Revolution*, Havana, 1980, p. 76.
14 *Manifesto of the New Jewel Movement*, St Georges, mimeo., 1973, p. 3.
15 A full summary was given in the *Barbados Advocate-News*, 27 January 1974.
16 C. Searle, 'Grenada's revolution: an interview with Bernard Coard', *Race and Class*, XXI, 2, 1979, p. 174.
17 *Manifesto of the New Jewel Movement*, p. 11.
18 W. R. Jacobs and B. Coard (eds.), *Independence for Grenada: Myth or Reality*, Trinidad, 1974, p. 143.
19 T. Thorndike, 'Maxi-crisis for mini-state', *The World Today*, XXX, 10, October 1974, pp. 440–1.
20 Sensing this, he played no part in regional affairs. He did, however, try to bolster his image by hosting the annual conference of the Organisation of American States in 1978. But this achieved nothing when it became known that one result was an arms deal with General Pinochet of Chile: see *Caribbean Contact*, November 1978.
21 *Grenada: The Route to Revolution*, p. 117.
22 The description is that of a CANA (Caribbean News Agency) report, reprinted in much of the Western media. See, for instance, *The Guardian*, 13 March 1982.

23 Information from interviews with NJM leaders, May 1979.
24 *New York Times*, 29 March 1979.
25 Gairy was particularly active among Cuban exiles in Miami and was promised support: see *Latin American Regional Report: Caribbean*, 23 November 1979.
26 R. E. Gonsalves, 'The importance of the Grenada revolution to the eastern Caribbean', *Bulletin of Eastern Caribbean Affairs*, V. March/April 1979, p. 8.
27 *Agreement between the Government of Cuba and the Government of Grenada*, St Georges, mimeo., n.d.
28 Information from interviews with NJM leadership, May 1979.
29 *Insight*, London, April 1982.
30 'Work Harder, Produce More, Build Grenada', *Report on the National Economy for 1981 and the Prospects for 1982*, Presented by Bro. B. Coard, St Georges, 1982, p. 64.
31 *Ibid.*, pp. 68–73.
32 *Ibid.*, pp. 26–33.
33 *World Bank Atlas, 1981*, Washington, DC, 1982.
34 Quoted in *One Month after the People's Revolution*, Address by Comrade Maurice Bishop, Revolutionary Prime Minister, 13 April 1979, St Georges, p. 4.
35 *Ibid.*, p. 5.
36 *Organise to Fight Destabilisation*, Address by Brother Maurice Bishop, St Georges, 8 May 1979.
37 *Free West Indian*, 26 July 1980.
38 See, for instance, *Caribbean Contact*, July 1980.
39 *Ibid.*, March 1981.
40 Quoted in H. S: Gill, 'The foreign policy of the revolution', *Bulletin of Eastern Caribbean Affairs*, VII, March/April 1981, p. 4.
41 *Forward Ever!*, p. 20.
42 *The New Jewel*, 13 March 1980 and 13 March 1981. In September 1980 there was a Soviet gift of over US$1 million worth of agricultural machinery, followed by a Bulgarian gift of ice-making plants to preserve fish; food, technical and medical aid from Hungary and the German Democratic Republic and scholarships by all the major Warsaw Pact states. In mid-1982 a major Soviet aid programme of over US$8 million was agreed. It constituted a trade agreement, a Soviet promise to reciprocate the placing of a resident Grenadian ambassador in Moscow with a Russian representative in St Georges and an inter-party accord between the NJM and the Communist Party of the Soviet Union. Cuba also agreed to further aid in 1982 for ancillary airport work.
43 *Caribbean Contact*, April 1982.
44 *A State of Danger in the Caribbean*, Testimony given by Assistant Secretary of State for Inter-American Affairs Thomas Enders, Washington, DC, 14 December 1981. See also *Aviation Week and Space Technology*, 21 December 1981, pp. 19–20.
45 *Caribbean Contact*, May 1981.
46 *Address by Prime Minister Bishop to certain Heads of Foreign*

Governments, International Organisations and prominent Political Personalities, St Georges, mimeo, 20 August 1981, p. 1.
47 *Barbados Advocate-News*, 31 August 1981.
48 *Insight*, April 1982. Later, however, the OECS decided not to reject new US aid if there were discriminatory terms against Grenada as long as the PRG continued to accept aid from third countries (the German Democratic Republic was specifically mentioned) on a bilateral basis.
49 *Barbados Advocate-News*, 15 November 1981. A total of US$2·5 million was eventually granted under the Lomé II arrangements.
50 *The New Jewel*, 11 April 1981.
51 *Feature Address by Prime Minister Maurice Bishop at the Official Launching of the Media Workers Association of Free Grenada*, St Georges, mimeo., 11 July 1981, p. 2.
52 *Latin American Regional Report: Caribbean*, 25 June 1981.
53 *Caribbean Contact*, August 1981.
54 *Barbados Advocate-News*, 14 March 1981.
55 *Is Freedom We Making*, pp. 41–3.
56 *Report on the National Economy*, pp. 76–7.
57 *Free West Indian*, 13 March 1981.
58 R. Hart, *In Defence of our Sovereignty*, Havana, September 1981, mimeo., n.d., Appendix 2, p. 6.

PART II

The regional level

Regional economic integration first came to the forefront of political thinking in the Commonwealth Caribbean in the late 1960s. With its implicit message of collective self-reliance, the appeal of the policy derived initially from the fact that the region was then just beginning to feel the cold draught of self-government. The various governments felt themselves to be under pressure on several fronts. They were concerned about the lack of economic development in the Commonwealth Caribbean. They were particularly anxious about the market prospects of their vital exports of primary products in the light of Britain's application to join the European Economic Community and her likely abandonment of Commonwealth preference. They were also alarmed by the way opportunities for migration out of the region to Britain, the United States and Canada were being reduced, since throughout the 1950s migration had contributed considerably to mitigating the effects of the region's overpopulation. In these circumstances, unity in adversity seemed to most Commonwealth Caribbean governments to be a slogan worth hanging on to. In the light of the region's bitter experience of federation a further experiment in political integration was unacceptable, but the way was open for the negotiation of a measure of regional economic integration.

The Caribbean Free Trade Association (CARIFTA) came into being on 1 August 1968, eliminating most tariff barriers between the various territories of the Commonwealth Caribbean region. For the purpose of the integration movement they were divided into two groups – the so-called More Developed Countries (MDCs) of Jamaica, Trinidad and Tobago, Guyana and Barbados, and the remaining Less Developed Countries (LDCs) of Belize and the other Eastern Caribbean islands. Although little progress was made in the next few years towards the declared aim of deepening the movement in the direction of full market integration, it was at least still in existence when, after 1970, the enhanced prospect of Britain joining the EEC again brought home to Caribbean governments the extent of their vulnerability to any disruption of the preferential trading ties historically established with Britain. They were forced to face up to the need to advance the integration movement to a new and deeper level if the economic well-being of the region was not to be seriously affected. Thus began the process of negotiating the various steps necessary to move beyond CARIFTA, a process which was eventually brought to a successful conclusion in July 1973 when the Treaty of Chaguaramas was signed, establishing the new Caribbean Community and Common Market (CARICOM).

The establishment of CARICOM constituted a considerable advance in the development of Caribbean integration. Its goals embrace three broad areas of co-operation. The first is the furtherance of regional economic integration by the establishment of a common market; the second is the expansion of functional co-operation in such fields as health, education, transport and meteorology; and the third is the co-ordination of foreign policy among the fully independent states of the Community. CARICOM was thus an ambitious venture, and in practice it has not proved easy to realise the aims set out in the Community Treaty.

From the outset CARICOM ran into problems. The ink was hardly dry on the signatures to the treaty when the full force of the international economic crisis which developed in 1973–74 and from which the world economy has yet to recover struck the incipient integration movement. The first sign of open discord came in April 1975 when the Prime Minister of Trinidad, Dr Williams, complained that the recent advances in Caribbean integration were being prejudiced by the way in which many of the impoverished member states of CARICOM were making bilateral economic arrangements on supplicant terms with wealthy Latin American countries. A row developed which caused the leaders of the Community to engage in public abuse of each other. In 1977 the dissension within the integration movement reached a still more serious level when Guyana and Jamaica were forced to restrict their imports even from CARICOM member states as a means of alleviating their desperate financial situation. By so dramatically exposing the fragility of the free-trade regime which was the mainstay of CARICOM, the affair constituted a major setback in the achievement of regional integration. It certainly clouded the atmosphere within the Community. Meetings of the CARICOM Council became infrequent, the Secretary General left, and the morale of all the staff of the Secretariat began to suffer. Most importantly, no progress was made towards important Community goals like the negotiation of a joint agreement on the terms on which foreign investment could enter the region or the establishment of regional industrial or agricultural programmes.

This last failing had been a target of radical criticism of Caribbean integration from the moment of its inception. The argument was that the initiation of regional free trade merely created a larger field of exploitation for foreign-owned transnational corporations and thus further entrenched the Commonwealth Caribbean's dependent position within the international economy. It was suggested that only by embracing what was called production integration, namely the establishment of regional industries and regional agricultural projects, could CARICOM help to build up the actual productive capacity of Commonwealth Caribbean economies and thereby reduce the region's long-standing economic underdevelopment.

Since the appointment of Dr Kurleigh King as Secretary General in November 1978 there has admittedly been a lessening of tension within the integration movement, and a sense of purpose has been restored to the work of the Secretariat. Nevertheless it is clear that CARICOM has reached a critical point in its development. The Community itself more or less admitted as much when in March 1980 the Council decided to appoint a prestigious team of regional experts to review the functioning of Caribbean integration and prepare a strategy for its improvement in the decade of the 1980s. The experts produced their report in 1981 and laid down a series of new goals for the Community, notably in the area of production integration.

The main problem the regional movement faces is, of course, politics. CARICOM has now to survive and prosper in a regional political context which is different in at least three important ways from the one in which it was conceived and established. Firstly, there no longer exists a genuine balance of power between the member states of the Community. An enormous gap has opened up between Trinidad, which is now unequivocally the dominant economy within CARICOM, and even the other MDCs. Only Trinidad can now give direction to the regional movement and, although in the last few years it has carped repeatedly at the inadequacies of CARICOM and the allegedly anti-integrative behaviour of some of its member states, it has not given any lead as to the direction in which it wants CARICOM to move.

Secondly, the consensus that existed between Commonwealth Caribbean governments in the 1960s about the appropriate strategy of development to follow has been replaced by what is termed 'ideological pluralism'. Much of the debate about the consequences for regional integration of the differing left-wing and right-wing stances adopted by governments within the region has been oversimplified and exaggerated, but it obviously is the case that member states of CARICOM now hold divergent views about the meaning of economic development and justify their policies by reference to conflicting ideologies. This is bound to make it harder to deepen the present level of regional integration.

Thirdly, the genuine *rapport* which existed between the heads of government of the main CARICOM territories during the period when the Community was being negotiated has given way to a number of rivalries and conflicts between regional leaders. The most bitter individual relationship was between Williams and Michael Manley, although that between Maurice Bishop and the Barbadian Prime Minister, Tom Adams, has recently run it close for depth of personal rancour. For CARICOM as a whole the result of these personality conflicts has been that the Heads of Government Conference, formally constituted as the sovereign body of the Caribbean Community, did not meet in full session from December 1975 to November 1982. This most recent gathering has raised new hopes, but cannot be expected to have fully restored the close personal relationships between the region's leaders which were such an important factor in the earlier stages of Caribbean integration.

In the face of these changes in the environment in which it has to operate, CARICOM has been immobilised and there are at the moment few real signs that the political conditions for its functioning can be recreated in such a way that the Community can be restored to health in the form in which it previously prospered. Indeed, it may be that the CARICOM model of integration has, at least for the foreseeable future, run out of steam. Nevertheless the Secretariat and the institutions of the Community are currently addressing themselves to the proposals made by the team of experts and are devoting particular attention to the matter of production integration. This has now become the testing ground for the future of regional integration in the Commonwealth Caribbean.

Anthony Payne

5 Regional industrial programming in CARICOM

Regional industrial programming has become one of the catch phrases of the Caribbean integration movement. It denotes the formal commitment of the Caribbean Community and Common Market to undertake to promote a process of industrial development between the member territories which would lead to the creation of more extensive production links within the Commonwealth Caribbean region. As such, it has been one of the main goals of CARICOM since the treaty was signed in 1973 and has subsequently been the focus of considerable attention on the part of regional policy-makers.

The thinking behind the policy of regional industrial programming is simple. It is based upon the belief that a coherent industrial policy for the Commonwealth Caribbean as a whole would encourage economies of large-scale production for the wider market, promote greater rationalisation and specialisation in different branches of existing industries and, above all, lead to the establishment of regional industrial complexes drawing upon the combined natural and technological resources of the entire area.[1] The ultimate objective is the creation of a stronger and more deeply integrated Caribbean economy better able to withstand the pressures of dependence on the international economic system.

Although not always couched in quite these terms, such thinking on the subject of regional industrial policy is not new. In different guises it has been present within the modern Caribbean integration movement since the movement's inception in the 1960s. This chapter discusses the various attempts that have been made to give effect to ideas of regional industrial development and the responses with which they have been met. It begins by describing the initial process of industrialisation in the Commonwealth Caribbean in the period after the end of the second world war.

Post-war industrialisation

The old colonial policy discouraged, if not forbade, the development of manufacturing industry in the West Indies and other British colonies. The Moyne Commission, which visited the Caribbean in 1938–39, considered that manufacturing was not an area of comparative advantage on account of the region's lack of fuel, raw materials and requisite skills and was adamant that West Indian governments should not engage in what were termed 'speculative industrial enterprises'.[2] The Commission did concede that certain bulky commodities, like cement, might be manufactured if the local consumer could be protected from unduly high prices consequent upon the anticipated inefficiencies. It suggested that British firms might be approached, to explore the possibility of establishing branch plants in the region. During the war shipping difficulties disrupted traditional supplies and stimulated the emergence in several of the islands of an embryonic industrial sector making a variety of foods, drinks and consumer items.[3] Official policy, however, seemed not to appreciate these developments. In his Economic Plan for Jamaica, produced in 1948, the Economic Adviser to the Comptroller of Development and Welfare, Dr Frederic Benham, still gave a very low emphasis to the possibility of building up manufacturing industry on the island.[4]

This orthodoxy was challenged most famously and successfully by Professor W. Arthur Lewis, the distinguished St Lucian economist, in two articles published in *The Caribbean Economic Review* in December 1949 and May 1950, and subsequently reprinted in pamphlet form by the Caribbean Commission.[5] In the first, Lewis reported in admiring terms on the nature of the industrial development which had been achieved in the previous few years in Puerto Rico, whilst in the second he argued the novel case for the rapid industrialisation of the West Indies. His thesis began with a consideration of the West Indian territories' long-standing economic role as primary commodity producers, but immediately departed from the traditional colonial perspective by demonstrating that agriculture was already unable to support the growing population of the islands and indeed could be made more efficient only if the numbers engaged in it were drastically reduced. This was the insight which in Lewis's mind brought the question of industrialisation to the fore. Contrary to widespread belief, he did not see industrialisation as an alternative to agriculture, but rather as 'an essential part of a programme for agricultural improvement'[6] which, by providing new jobs, would take surplus labour off the land. Industrialisation thus emerged in the Caribbean out of the back pocket, as it were, of agriculture.

From this point of departure Lewis proceeded to set out a policy of industrialisation for the Commonwealth Caribbean designed to overcome the dual problems of resources and markets. In the first matter, the secret of success was to specialise in those areas of manufacturing for which the region's resources were most appropriate. The region was short of capital, industrial power was expensive and the available raw material base limited, but wage rates were low by the standards of the developed world. Working on this basis, he was able to draw up a list of 'most favourable industries' which the islands could establish fairly cheaply.[7] Many was based not on the use of local raw materials but on the processing of imported inputs. As for markets for these products, the fact that the territories individually were too poor and too small necessitated the establishment of a regional customs union as an essential prelude to any vigorous policy of industrialisation. 'There are few industries,' Lewis wrote, 'in which the market of all the islands together could support more than a single factory, working on a reasonably economic scale, and manufacturers are naturally reluctant to establish unless they can treat the whole market as a single unit. It is risky to establish in Trinidad, if five years later the Jamaica Government is going to give tariff protection to a new Jamaican plant.'[8] To Lewis, though, even this was not enough: regional import substitution would account for only a small part of the industrial output necessary to generate full employment. Export-promotion industrialisation was the main requirement.

The final part of Lewis's argument addressed itself to the difficulties faced by the Commonwealth Caribbean in endeavouring to break into established export markets. Here the key was not to attempt to establish new external trade channels, but rather to persuade manufacturers who were already selling in overseas markets to locate their plants actually in the Caribbean. This in turn required a number of governmental initiatives in direct contravention of the traditional *laissez-faire* economic philosophy of the region. For Lewis, industrialisation was like 'a snowball': once started, it would move of its own momentum and get bigger and bigger as it went along. To attract foreign manufacturers and thus get the snowball rolling, he recommended the implementation of a package of investment incentives modelled upon recent Puerto Rican experience – the provision of basic infrastructural services, the establishment of special Industrial Development Corporations and Development Banks, the enactment of appropriate policies of protection via tariffs and quotas and the proclamation of tax 'holidays' for newly established industries, including, in particular, the remission of taxes on imported raw materials and machinery.[9] Eventually, in Lewis's view, the inflow of foreign investment would produce sufficient profits, generate sufficient local savings and transmit sufficient skills to local people to set in motion self-sustaining industrial growth.

Lewis's prescription for industrial development had an immediate impact on newly-emerging Commonwealth Caribbean governments anxious to secure economic achievements with which to impress their electorates once universal adult suffrage began to be introduced into the region from 1944 onwards. A period of what Lewis himself referred to as 'wooing and fawning upon'[10] foreign capitalists followed the publication and dissemination of his analysis. The so-called Aid to Pioneer Industries Ordinance was passed in Trinidad in 1950 and the Jamaican Industrial Development Corporation set up in 1952. Before long all the MDCs and some of the LDCs had established the institutional and legal apparatus attendant upon this policy of 'industrialisation by invitation', as it was soon accurately, if derisively, dubbed.[11] Foreign capital responded to the supine posture of the region's governments and flowed into the area in massive quantities, bringing in its wake a number of highly visible manufacturing industries. By 1967 manufacturing contributed 15% of the gross domestic product of Jamaica and 16% in the case of Trinidad, whilst the figures for Guyana and Barbados were 13% and 9% respectively.[12] The LDCs had done less well in statistical terms, but nevertheless light industrial estates were beginning to appear even on the smaller islands.

At first sight, therefore, the Lewis strategy of industrialisation appears to have been strikingly successful, but it was not in fact properly implemented. Instead of being fuelled by export promotion, most of the industrial growth achieved in the Commonwealth Caribbean during the 1950s and '60s was oriented towards import substitution and, with the collapse of the West Indies Federation in 1962 after just four years' existence, concentration was generally upon the insular rather than the regional market. During this period the larger West Indian territories came to produce for themselves a variety of manufactured goods – most typically, shoes, cosmetics, garments, furniture, paint, aerated drinks, beer, cigarettes, soap, and latterly products like woven textiles, cement and bottles. What did not emerge as extensively as Lewis had originally hoped was a thriving 'enclave' sector geared to overseas export markets. In Jamaica, for example, even though a special Export Industry Encouragement Law was passed in 1956, only forty out of 130 firms which started production under the incentive laws between 1950 and 1964 were enclaves.[13] The relative lack of success of this part of the strategy in Jamaica was consistent with the experience of the rest of the Commonwealth Caribbean and was reflected in the fact that exports of manufactured goods represented only 4·4% of manufacturing gross domestic product for the region as a whole in 1970.[14] Quite apart from the considerable import intensity in terms of unassembled intermediate inputs and the high degree of foreign ownership of industrial plant permitted and indeed encouraged by incentive-fostered industrialisation, the emphasis on insular import substitution, rather than export promotion, in the context of small domestic markets inevitably limited markedly the transformative impact that the emergence of an industrial sector was able to have on the Commonwealth Caribbean economy as a whole. It was largely as a consequence of these limitations that industrial policy in the area acquired the regional dimension Lewis had proposed at the outset.

The experience of CARIFTA

Regional economic integration in the Commonwealth Caribbean was conceived fundamentally as a mechanism for correcting some of the flaws which had beset the strategy of 'industrialisation by invitation'. Lewis's heir as architect of the next phase of the region's industrial development was William Demas, who in 1965, as head of the Economic Planning Division of the Trinidad government, published an original and highly influential analysis of the condition of the Commonwealth Caribbean economy.[15] Demas pointed to two important weaknesses in the postwar pattern of industrial growth in the region. Firstly, the level of unemployment and underemployment, which had been the main motive force behind Lewis's thinking, was still high and, so far from decreasing over the previous decade, was thought actually to have grown. The high wage rates paid in the new manufacturing sector had the effect of raising the reserve price of labour and thus of persuading people to give up low-paid agricultural employment in order to join the ranks of the urban unemployed. At the same time, most of the imported technology used in the industrial sector had turned out to be highly capital-intensive and unsuited to the special needs of the labour-surplus economies into which it had been introduced. For example, the 146 industries which, by the end of 1965, had been established in Jamaica under the incentive legislation programme had provided a mere 9,000 jobs,[16] whilst in Trinidad the employment impact of industrialisation had proved to be equally disappointing: only 4,666 jobs were made available by ninety-nine new industries between 1950 and 1963.[17] Many more were needed, in view of the continuing increase in the population of the region and the closure of emigration outlets in the 1960s, just to prevent the unemployment level from worsening. Secondly, the use made of local West Indian resources in the process of industrialisation had been negligible. Demas noted that foreign investors in the manufacturing sector had preferred, on the whole, to locate within the Commonwealth Caribbean no more of the production process than was necessary to be awarded the tax incentives. As a result, the industries that had grown up were often no more than 'screwdriver' final assembly operations which had failed to establish close linkages with the rest of the local economy.

Demas sought salvation from these problems in the regional connection. He argued that the resulting failure of Commonwealth Caribbean states to transform their economies during this period of industrial growth was to be explained largely by their small size, defined in terms of both land area and population.[18] The smallness of the domestic market, he reasoned, imposed sharp limits on the process of import-substitution industrialisation and thus removed the option of balanced growth, incorporating a roughly equal mixture of export promotion and import substitution, a goal which could only really be attained by large continental countries. For the small Caribbean territories this path of development could most satisfactorily be secured by economic integration with neighbouring underdeveloped countries. Demas anticipated that integration would not only promote industrial growth by making possible the elimination of excess capacity in existing manufacturing industry, but also, and more important, by stimulating investment in new industries which would become economically feasible for the first time in the Commonwealth Caribbean on the basis of the expanded market. He did stress that 'integration may often not remove the necessity to seek export markets outside the region',[19] but the lack of attention given to this facet of industrialisation showed how far his thinking had moved from that of Lewis. No one was in any doubt

Industrial programming in CARICOM

that, in Demas's mind, the key to the strategy of economic integration was the pursuit of import substitution on a regional basis. As he himself put it, 'the creation of an economic region can mean that the development pattern for the region *as a whole* can approximate more to import-substitution − although from the point of view of individual member countries there will still be a large volume of "exports" to and "imports" from other countries'.[20]

CARIFTA, when it was established in 1968, was modelled directly upon Demas's thinking. From the outset the new regime liberalised the great bulk of intra-regional trade, thereby extending the size of the 'home' market for the region's various light industries. CARIFTA had its political difficulties, but in its main economic function it worked much as planned. Prior to CARIFTA, intra-area trade was made up primarily of petroleum products from Trinidad, fertilisers, some chemicals and cement from Trinidad and Jamaica, rice from Guyana and root crops from the Leeward and Windward Islands. From the beginning of regional free trade the trade flows revealed a relative decline in the trade in petroleum and petroleum products, an increase in food and beverages and, above all, a substantial expansion in the trade in manufactured goods, i.e. goods coming under SITC sections 5–8. According to figures gathered together by the Economic Commission for Latin America, intra-CARIFTA trade in manufactures increased between 1967 and 1971 at nearly twice the average rate of growth for all items.[21] Indeed, by 1972 Trinidad was consigning 34% of its exports of manufactured goods to other CARIFTA countries, Jamaica 49%, Barbados 40% and Guyana 66% − all very sizeable proportions.[22]

The weakness underlying these otherwise impressive figures is that, because of the particular type of industrial development pursued by the region, the 'value added' in the production of manufactured goods is relatively small − only about 50% in the upper reaches and, in many cases, considerably below that level. This sort of trade was specifically underwritten within the CARIFTA Agreement by what was known as the Basic Materials List.[23] For the purpose of defining rules for area tariff treatment, this conferred local origin on materials which are not, and never have been, produced in the region, but which were needed to sustain previously established light manufacturing industries. Regional technocrats have long recognised this deficiency in regional trading arrangements, but have found it difficult to persuade the governments to take action.[24] In consequence, as Demas himself was eventually forced to admit, a great deal of intra-Commonwealth Caribbean trade 'from a strictly economic point of view may not be all that beneficial to the member countries who are exporting'.[25]

By the beginning of the 1970s a growing awareness of the limitations of CARIFTA began to develop, even among the officials of the integration movement. This gave rise, in turn, to a deeper questioning of the original industrialisation strategy, of which regional free trade was, after all, no more than a neat extension. Indeed, it was the Commonwealth Caribbean Regional Secretariat, the administrative agency of the integration movement, which, in 1972, offered this analysis of the region's pattern of industrial growth:

> the rapid growth of the manufacturing sector which has taken place in the last decade or two in the four independent countries has been accompanied by certain undesirable features: too heavy a dependence on foreign capital, foreign technology and foreign inputs (that is, raw materials and components); the excessive capital-intensity (and therefore the limited impact on employment) of the foreign technologies used; the creation of insufficient 'linkages' between this sector and other sectors of the economy (particularly agriculture); a drain abroad of profits, dividends, interest, royalties, licence fees and management charges because of heavy dependence on foreign capital and foreign enterprise; and insufficient expansion of exports of manufactures to countries in the outside world — a very important objective for small countries such as those in the Caribbean.[26]

This added up to a damning indictment of the first two stages of the Commonwealth Caribbean's industrial experiment and led to the adoption of a new and deeper-seated approach to regional industrial development when CARICOM replaced CARIFTA as the agent of regional economic integration in 1973. Significantly, this new approach was a partial response to the radical critique of West Indian economic development which had emerged with some force during the 1960s.

The concept of production integration

Haunting the whole existence of CARIFTA, almost as a ghost, was an alternative conception of regional economic integration developed by a group of economists at the University of the West Indies (UWI). In a series of studies on the theme of integration, the most comprehensive and the most discussed was *The Dynamics of West Indian Economic Integration* by Havelock Brewster and Clive Thomas.[27] The main thesis of this work was that a free-trade area and customs union, on their own, did not constitute an adequate structure for procuring the gains from integration. Sufficient conditions were established only when this approach was combined with

> a functional and sectoral approach which explicitly allows for the introduction of planning techniques in an effort to ensure the positive development of certain agreed areas of economic activity in the region. In other words integration in the West Indies should not be limited to those conditions which govern the exchange of goods, but should also include in its perspective the integrated production of goods.[28]

The emphasis of the study was therefore on the possibilities of production integration, 'taking place from the basic resource inputs of the particular sector to the disposition and sale of final output yielded by the sector'.[29]

Brewster and Thomas elaborated their approach in the greatest detail in respect of industry. They argued that the resource base of the region contained many of the basic materials required to develop an advanced industrial sector, but that such a structure could be created in the Commonwealth Caribbean only by pooling the resources of the different countries in order to establish regional industrial complexes or 'integration industries'. They sought to identify the areas with the highest degree of development potential and, at the end of their research, were able to conclude that, by following their recommendations, the Commonwealth Caribbean could look forward to the establishment by 1975 of one steel mill based on imported scrap, one plant producing paper other than newsprint, one synthetic nylon factory, one sheet and plate glass factory, one caustic soda plant, one factory producing general-purpose rubber and one car factory, building vehicles with 60% value added from regional components.[30]

Advocacy of the concept of production integration added a new dimension to the debate about regional industrial policy in the Caribbean. Even when the two authors appeared to talk the same language as Demas, in stating that the immediate purpose of economic integration was to promote regional import substitution, they usually meant different things – in this instance, interpreting regional import substitution not just as a means of reducing the level of imports but as a prime feature of economic development signifying the growth and structural transformation of regional productive capacity. Economic integration was not to be judged primarily by the extent to which items previously imported were subsequently manufactured within the region, but by the extent to which it reduced the wide disparity between the structure of domestic demand and the structure of domestic resource used forced upon the Caribbean economy by the fact of its economic dependence. In short, Brewster and Thomas's priority was to see the region consuming more of what it could produce, rather than producing more of what it wanted to consume – a fine but important distinction.

Achievement of this end, in their view, demanded a number of fundamental breaks with the general economic and industrial policy pursued throughout the Commonwealth Caribbean in the immediate past. They argued, firstly, that final-assembly manufacturing could never lead to the economic transformation of the region, and that widespread emphasis on this sort of 'horizontal' industrialisation should be replaced by a greater concern for 'vertical' industrialisation of the type they proposed; secondly, that there was 'a very strong case for the public ownership and development'[31] of integration industries in order to prevent their monopoly of the regional market being put to private advantage; and thirdly, that there was an urgent need to reorganise those areas of the economy in which the other major basic resources of the region, such as bauxite, bananas and sugar, were integrated into the operations of multinational corporations based in metropolitan areas. Brewster and Thomas did not direct their attention to this last area of policy,[32] but it was integral to their wider vision of regional integration as the central feature of a radical social and economic transformation of the Commonwealth Caribbean.

In the event, the concept of production integration, as advanced by Brewster and Thomas, was not taken up by the region's governments. It was given a perfunctory hearing at the vital conference of officials in Georgetown in August 1967 which effectively laid the groundwork for CARIFTA but was never considered seriously. The reason is straightforward: the strategy challenged many of the assumptions and demanded the overthrow of many of the policy shibboleths of the previous decade, not least in the field of industrial policy. The business sector was naturally alarmed at the emphasis on the public ownership of integration industries, and their representatives at the Georgetown conference missed no opportunity to propagate the merits of private capital, both local and foreign, in engineering economic development. For their part, the governments of the region were also too conservative in their attitude to political economy to risk embarking upon the road outlined by Brewster and Thomas.

It was, of course, claimed that something of their approach was salvaged within the adopted format of Caribbean integration.[33] Reproduced within the CARIFTA Agreement, as Annex A, was the text of a resolution passed by the fourth Heads of Caribbean Governments Conference in Barbados in October 1967. This set out some of the future goals of the integration movement and contained as one of the items the following statement:

> The principle should be accepted that certain industries may require for their economic operation the whole or a large part of the entire regional market protected by a common external tariff or other suitable instrument. The location of such industries and the criteria to be applied in respect thereof, as well as the implementation of the principle accepted above, should be the subject of immediate study.[34]

In the context of a free-trade movement this read rather like a token commitment, and certainly the promised study was pursued with no great zeal. Eventually, in August 1969, a team of consultants from the United Nations Industrial Development Organisation submitted a report and four preliminary sectoral studies covering textiles, food processing, chemicals and pulp and paper to member governments[35] – only to see them disappear, almost without trace, into the political stalemate into which Caribbean integration had fallen by early 1970. Another report focusing upon industrial development possibilities within the LDCs was produced by the Economist Intelligence Unit in 1972,[36] but by that time CARIFTA was in the process of being transformed into the more ambitious Caribbean Community and Common Market.

CARICOM: Article 46 of the Common Market

The Caribbean Community Treaty contained among its many provisions a new commitment to regional industrial policy-making which seemed to meet some of the criticisms levelled at the approach adopted in CARIFTA. Article 46 of the Common Market Annex to the Treaty read as follows:

> Member States undertake to promote a process of industrial development through industrial programming aimed at achieving the following objectives:
>
> (a) the greater utilisation of the raw materials of the Common Market;
> (b) the creation of production linkages both within and between the national economies of the Common Market;
> (c) to minimise product differentiation and achieve economies of large scale production, consistent with the limitations of market size;

(d) the encouragement of greater efficiency in industrial production;
(e) the promotion of exports to markets both within and outside the Common Market;
(f) an equitable distribution of the benefits of industrialisation paying particular attention to the need to locate more industries in the Less Developed States.[37]

It is hard to know how to evaluate such a commitment. On paper it provides adequately for the adoption of schemes of regional industrial programming and also has the virtue of flexibility. The Regional Secretariat apparently held the view that 'generally speaking, the procedures for establishing regionally integrated industries should not be hamstrung by any formal Protocol or Agreement on a Regime for regionally integrated industries, as has bogged down the process in several other integration groups'.[38] On the other hand, from the perspective of the Brewster and Thomas approach, Article 46 has a distinctly woolly flavour. The only answer is to see whether a more substantive embrace of industrial programming has subsequently been negotiated between the member governments.

For a time, at the end of 1974 and the beginning of 1975, the formulation of plans for a number of regional production projects gave genuine grounds for believing that this part of the treaty was indeed being implemented in a meaningful way. The most spectacular of these projects was announced in June 1974, when the governments of Guyana, Jamaica and Trinidad made public their intention to build two aluminium smelters, one in Trinidad and one in Guyana. The Trinidad smelter, which it was hoped would be established by 1977, was to derive its power from the island's natural gas and was to be jointly owned by the three governments, with the Trinidad government holding 34% of the equity and the Jamaican and Guyanese governments 33% each, whilst the Guyanese smelter, which was not due to be completed until 1980, was to use hydro-electric power and was to be owned 52% by the Guyanese government and 24% each by the other two. It promised to be a major step towards the integration of regional production in that it aspired to utilise in combination some of the region's most valuable raw material resources, and it certainly led to a bullish phase of comment and analysis about the future prospects of the Caribbean Community.[39] Sustaining this mood of optimism was the emergence of a less spectactular, but nevertheless significant, proposal for a cement plant, located in Barbados but based on a joint enterprise between Barbados and Guyana, the latter taking clinker cement from a Barbados factory for final processing in Guyana, and the announcement of an industrial allocation scheme between those CARICOM LDCs that belonged to the Eastern Caribbean Common Market (ECCM). This grew out of the Economist Intelligence Unit survey of industrial opportunities open to the LDCs, first published in 1972, and endeavoured to avoid wasteful duplication in the implementation of its proposals by allocating some thirty-five potential new industries on an agreed and reasonably equitable basis between the seven ECCM states.[40]

However, events since 1974–75 have tended to belie the hopes of that time as far as regional production integration is concerned. The smelter project became embroiled in an acrimonious intergovernmental dispute between Trinidad and Jamaica about the role to be played in the regional economy by powers outside the Commonwealth Caribbean. The late Trinidadian Prime Minister, Dr Eric Williams, was angered by an economic co-operation agreement signed in April 1975 between the governments of Venezuela and Jamaica, in which the latter agreed to supply Venezuela with considerable quantities of bauxite and alumina to fuel a planned increase in Venezuelan smelter capacity. In Williams's view it was 'simply not possible' to view Venezuelan policy 'as anything but a calculated attack' upon the CARICOM scheme[41] – an attack, moreover, to which one of Trinidad's regional colleagues was a party. His patience exhausted, Williams announced his government's withdrawal from the joint scheme in bitter and portentous words. 'My friends,' he said to his party convention, 'one man can only take so much, and I have had enough. To smelt or not to smelt, no big thing as there is no shortage of claims on our gas.'[42] The whole project has accordingly been abandoned, Trinidad subsequently declaring its intention to 'go it alone' with the building of a national smelter. There has been similar lack of progress with the cement plant, partly owing to technical considerations,[43] which leaves the ECCM allocation scheme as the region's only practical achievement to date in the field of industrial programming. Here too no more than seven or eight of the proposed industries have actually come on stream.

At the formal CARICOM level in relation to the specific commitment of Article 46, there has been some activity behind the scenes but little, as yet, to show for it. To oversee the implementation of the Treaty's aims in this area of policy the Ninth Heads of Government Conference, held in St Lucia in July 1974 shortly after the Caribbean Community had come into being, decided to establish a Standing

Committee of Ministers responsible for Industry, advised by a working party of industrial planners, as one of the so-called 'institutions' of the Community.[44] This committee convened for its inaugural session in 1975 but then did not meet for another six years. In the absence of a lead from Ministers, work on regional industrial programming has been confined to a plethora of technical committees, study groups and working parties, staffed by officials of the Regional Secretariat, the ECCM Secretariat, the Caribbean Development Bank and a variety of outside and United Nations bodies.

The latest exercise in this preparatory work does, however, seem to be the most thorough and serious yet conducted. In 1979 the CARICOM Secretariat obtained the approval of the Caribbean Development Bank for the financing of a study by a team of outside consultants. This had as its objective the preparation of a priority list of sub-sectors of the regional economy suitable for industrial programming, together with an appropriate list of specific industries for pre-investment analysis within the region. The terms of reference supplied by the Secretariat emphasised that regional industrial programming was expected 'to create new productive capacity serving both the regional or extraregional (or export) market, and should serve the ultimate goal of strengthening and integrating the productive structure of the respective member countries of CARICOM with a view to making both the national and regional productive structures more competitive in terms of world trade.'[45] The report was completed in April 1980 and constitutes the present basis of the Community's regional industrial strategy.

As the report was conceived only as the first step in what was intended to be a continuous planning process, the consultants set out at some length the analytical framework which had guided their recommendations. It consisted basically of three phases of analysis. The first sought to isolate particular clusters of interrelated industrial activities which could realistically be developed in the light of the priorities of regional demand and resource availability. Considerable emphasis was thus placed on the selection of projects with potential for the establishment of forward and backward linkages with other parts of the economy. At the same time consultation with governments and regional institutions was undertaken on the assumption that regional industrial programming projects which had already aroused interest would be easier to implement in the short term. The second phase involved the correlation of these two procedures and the identification of an appropriate group of specific projects for each cluster of industry. Finally, in the third phase the specific projects were classified into short, medium and long-term possibilities and preliminary pre-feasibility studies undertaken on the former category.[46] According to the consultants, such a planning framework had both the technical capacity to relate particular industrial projects to wide planning perspectives and the political flexibility to adapt to changing governmental priorities.

At any rate, on the basis of this framework, the report identified seven clusters of industrial activity as priorities for regional programming: food, textiles and clothing, wood and furniture, leather and footwear, shelter, petrochemicals and paper and pulp.[47] Some were clearly related to the 'basic needs' of the population; some others depended entirely on the development of regional resources. Most fell into the category of CARICOM imports. From these clusters over twenty short-term projects were specified and pre-feasibility studies undertaken. Since the report was presented further studies have been set in motion in the selected areas and a new working party of officials established to supervise the exercise. In February–March 1982 the Standing Committee of Ministers responsible for Industry met again for only the second time and made an attempt actually to allocate a number of industrial activities between member states of CARICOM. Some of its ideas were put to a Common Market Council meeting in July 1982 but were referred back to the Industry Ministers for yet further study.

Conclusion

These current plans do at least inject a note of qualification into what has otherwise been a record of almost complete failure to bring about a successful programme of regional industrial production in the Caribbean. A decade and a half after Brewster and Thomas published their formidable tome, nearly a decade now since the Caribbean Community Treaty was signed, not a single 'integration industry' exists in the Commonwealth Caribbean. It can readily be admitted that the deterioration in the condition of the world economy after 1973 has not been helpful, especially as it coincided with the moment when the early, and relatively easy, phase of import substitution industrialisation in the region first came in need of 'deepening'. Nevertheless, the 'Group of Experts' appointed in 1980 to review the record of Caribbean integration and chart its course for the future were more than justified in making the bland observation in their report that 'efforts to programme industrial production on a regional basis have been slow and disappointing'.[48]

Part of the explanation is, of course, that national strategies of industrialisation have been pursued with more vigour in the Commonwealth Caribbean throughout this period than regional programmes. Manufacturing industry in the region is, as a result, more widespread; yet, with the exception of Jamaica, Guyana, Barbados and Trinidad, it still accounts for an insignificant part of overall economic activity. Even in the four MDCs its role is not critical and has, in fact, stagnated in relative terms over the last decade. In 1980 the contribution of manufacturing to national GDP in Jamaica was 15·2%, compared to 15·7% in 1970; in Guyana 12·2%, exactly the same share as in 1970; in Barbados 10·9%, compared to 10·3% in 1973; and in Trinidad 7·2%, compared to 9·5% in 1973.[49] These figures indicate the existence of protectionist industrial interests which do not necessarily look favourably upon CARICOM schemes of industrial planning, but at the same time they suggest that if industrialisation is to fulfil its potential as a means of reducing the economic dependence of the Commonwealth Caribbean upon the outside world a substantial degree of regional industrial programming will have to be established. Although the position in this respect is perhaps more hopeful than for some time, several major problems remain to be resolved.

In the first place, as recent Ministerial meetings have shown, agreement has yet to be reached upon the allocation between the various member states of the chosen projects for industrial programming. None of the studies – not even that of Brewster and Thomas – has properly addressed itself to this thorny problem. No doubt some opportunity exists for trade-offs between particular countries, but it is as well to remember that the issue of the unequal distribution of benefits within the integration movement, in particular as between the MDCs and the LDCs, has been a highly political matter ever since preliminary negotiations about the establishment of CARIFTA were begun in the mid-1960s. In addition, Trinidad's possession of ample reserves of oil and natural gas and its consequent emergence since 1973 as a relative 'pole of growth' in the regional economy means that a heavily directed approach to the allocation question will be required if Trinidad is not to become even more obviously (and uncomfortably from the point of view of furthering integration) a super-MDC within CARICOM.

Secondly, greater efforts need to be made to encourage the

movement of capital, management and skilled labour within CARICOM and to provide easier rights of establishment within the region for West Indians wanting to set up economic enterprises outside their home country. The relevant provisions of the CARICOM Treaty in this area of policy were so weak as to be meaningless.[50] Some steps were taken to ameliorate the situation with the signing of an agreement setting up the so-called CARICOM Enterprise Regime, which is designed to facilitate the establishment of companies substantially owned by nationals of at least two member states. However, although a majority of CARICOM states signed the agreement as long ago as 1977, the recalcitrance of some has meant that it has not yet been brought into force. Among the most urgent recommendations of the 'Group of Experts' was that action be taken to remedy this deficiency and to award CARICOM Enterprise status automatically to any firm involved in implementing projects within the regional industrial programme.[51]

Thirdly, the provision of capital to finance regional industrial programming has to be considered. Hardly any of the CARICOM governments have sufficient funds to sustain much of a public sector involvement, and in any case a number of statements emanating from the Regional Secretariat over the past two or three years indicate that a key role is to be given to private business interests in the next phase of the integration movement. The problem is that local capitalism in the Commonwealth Caribbean is still predominantly oriented towards trade, rather than production. This leaves open the part to be played by foreign capital and multinational corporations. If such participation is either sought or careful restrictions on its involvement not established, especially in the matter of ownership of enterprises, the danger will exist that much of the potential gain from regional industrial programming will accrue outside the region and that this opportunity for resisting dependence will in fact lead to its further entrenchment.

Fourthly, an approach still needs to be developed to secure the promotion of industrial exports outside the region. A generation after the publication of Arthur Lewis's prescription for the industrialisation of the West Indies, this matter remains on the policy agenda. The clusters of industrial activity recently proposed by the team of consultants continue to be predominantly of an import-substituting nature. In view of the limited capital and entrepreneurial skills available in the Caribbean, there is an equally clear case for the regional programming of export industries, involving the utilisation of

CARICOM as a negotiating agent for markets overseas, especially within the framework of opportunities offered by the Lomé Convention in Europe and the Generalised System of Preferences in the USA and Canada. Regional policy-makers will sooner or later have to turn their minds to this vital aspect of industrial development. Its omission is critical.

Lastly, underlying all the other factors, there is the question of whether the politicians at the helm of the Caribbean Community have the political will to confront all these difficulties and guide the integration movement towards a serious embrace of the concept of production integration. It is widely accepted that a higher degree of intergovernmental political commitment is required to achieve success in this sphere than was necessary to effect intra-regional trade liberalisation. The challenge comes, moreover, at a time when CARICOM has still not fully recovered from the severe disruption wrought in its own institutions and in the politics of the region generally by the international economic disorder of recent years. In these circumstances it is not easy to see from where will come the necessary political sustenance to achieve meaningful regional industrial programming. The problem is therefore very serious as far as the whole future potential of Caribbean integration is concerned. As the 'Group of Experts' rightly observed, 'to a very considerable extent the success of CARICOM hinges on the extent to which the operation of the CARICOM arrangements leads to increased industrial production and employment in the countries of the region'.[52]

Notes

The author would like to thank the Nuffield Foundation for the provision of funds to finance fieldwork in the Commonwealth Caribbean in 1980.

1. For a fuller statement of these goals see Caribbean Community Secretariat, *The Caribbean Community: A Guide*, Georgetown, 1973, pp. 34–6.
2. *Report of the West Indian Royal Commission 1938/9*, Cmd 6607, London, 1945, p. 443, para. 46.
3. See R. Gallotti, *Industrial Development in the British Territories of the Caribbean*, Report prepared by the British member of the Industrial Survey Panel appointed by the Caribbean Commission, Port of Spain, 1948. For a detailed discussion of this process in one particular Commonwealth Caribbean country, see also E. Carrington, 'Trinidad's post-war economy, 1945–50' in N. Girvan and O. Jefferson (eds.), *Readings in the Political Economy of the Caribbean*, Kingston, 1971, pp. 122–32.

4 Frederic Benham, *Economic Plan for Jamaica*, Report prepared by the Economic Adviser to the Comptroller of Development and Welfare, London, 1948.
5 W. A. Lewis, 'Industrial development in Puerto Rico', *Caribbean Economic Review*, 1, 1949, pp. 153–76, and 'The industrialisation of the British West Indies', *Caribbean Economic Review*, 11, 1950, pp. 1–61. The pamphlet, which received much wider attention, appeared as *Industrial Development in the Caribbean*, Port-of-Spain, 1951.
6 Lewis, *Caribbean Economic Review*, 1950, p. 7.
7 *Ibid.*, pp. 26–7.
8 *Ibid.*, p. 30.
9 For the details see *ibid.*, pp. 37–53.
10 *Ibid.*, p. 38.
11 By the Group of New World economists at the University of the West Indies. See Girvan and Jefferson, *Readings*, Introduction, p. 1.
12 Commonwealth Caribbean Regional Secretariat, *CARIFTA and the New Caribbean*, Georgetown, 1971, p. 10. Manufacturing is defined to include sugar-milling operations, although they are obviously closely connected with sugar cane production in the agricultural sector. To this degree the figures exaggerate the extent to which new non-traditional industries were established in this period.
13 S. De Castro, *Tax Holidays for Industry: Why we have to Abolish them and How to do it*, New World Pamphlet No. 8, Kingston, 1973, p. 5.
14 S. E. Chernick, *The Commonwealth Caribbean: the Integration Experience*, Washington, 1978, p. 181.
15 W. G. Demas, *The Economics of Development in Small Countries with Special Reference to the Caribbean*, Montreal, 1965.
16 O. Jefferson, 'Some aspects of the post-war economic development of Jamaica' in Girvan and Jefferson, *Readings*, p. 112.
17 E. Carrington, 'Industrialisation by invitation in Trinidad since 1950' in *ibid.*, p. 144.
18 Demas, *The Economics of Development*, pp. 21–2.
19 *Ibid.*, p. 89.
20 *Ibid.*, Demas's emphasis.
21 Economic Commission for Latin America, 'The impact of the Caribbean Free Trade Association (CARIFTA)', *Economic Bulletin for Latin America*, XVIII, 1973, p. 144.
22 *Ibid.*
23 *Agreement establishing the Caribbean Free Trade Association*, Georgetown, 1968, Annex C Schedule – Basic Materials List. The list contained seventy-three materials.
24 Under the Caribbean Community Treaty the arrangement was extended and widened to include even more items. Proposals to tighten up these regulations were not implemented until June 1981.
25 W. G. Demas, 'The Caribbean Community and the Caribbean Development Bank', speech delivered at a Seminar on Management in the Caribbean, Port-of-Spain, 2 December 1975, mimeo, p. 5.
26 Commonwealth Caribbean Regional Secretariat, *From CARIFTA to*

Caribbean Community, Georgetown, 1972, p. 12.
27 H. Brewster and C. Y. Thomas, *The Dynamics of West Indian Economic Integration*, Kingston, 1967.
28 *Ibid.*, p. 19.
29 *Ibid.*, p. 25.
30 For the detailed arguments in support of these recommendations see *ibid.*, pp. 132–284.
31 *Ibid.*, p. 33.
32 This was covered in studies produced at the same time by other economists from the University. They included G. L. Beckford, *The West Indian Banana Industry*, Kingston, 1967; G. L. Beckford and M. H. Guscott, *Intra-Caribbean Agricultural Trade*, Kingston, 1967; and N. Girvan, *The Caribbean Bauxite Industry*, Kingston, 1967.
33 W. G. Demas, 'CARIFTA and the University studies', *The Nation*, 27 June 1969.
34 *Agreement establishing the Caribbean Free Trade Association*, Georgetown, 1968, Annex A.
35 United Nations Industrial Development Organisation, *The Establishment of Integrated Industries in the Caribbean and the Location of Industries in the Less Developed Countries*, New York, 1969.
36 Economist Intelligence Unit, *Eastern Caribbean and British Honduras Industrial Survey, Final Report*, London, 1972.
37 *Treaty establishing the Caribbean Community*, Chaguaramas, 4 July 1973, Annex – The Caribbean Common Market, Article 46(1).
38 Commonwealth Caribbean Regional Secretariat, *From CARIFTA to Caribbean Community*, p. 87.
39 See, for example, W. G. Demas, *West Indian Nationhood and Caribbean Integration*, Bridgetown, 1974, p. 65.
40 See Caribbean Development Bank, *Annual Report 1975*, Bridgetown, 1976, p. 16..
41 *Trinidad Guardian*, 16 June 1975.
42 *Ibid.*
43 See W. G. Demas, *Creating National and Regional Linkages in Production* (address to the eighth Annual Meeting of the Board of Governors of the Caribbean Development Bank), Bridgetown, 1978, p. 32.
44 *Final Communiqué of the Ninth Heads of Governments Conference*, Castries, 1974.
45 CEGIR, 'A Study on Regional Industrial Programming in the CARICOM Countries. Executive Summary', mimeo, 1980, p. 1.
46 See *ibid.*, pp. 8–11.
47 *Ibid.*, p. 16.
48 The Group of Caribbean Experts, *The Caribbean Community in the 1980s*, Georgetown, 1981, p. 50.
49 Economic Commission for Latin America, Office for the Caribbean, *Economic Activity 1980 in Caribbean Countries*, CEPAL/CARIB 81/10, Port-of-Spain, 1981.
50 See Demas, *West Indian Nationhood*, p. 63.

51 Group of Caribbean Experts, *The Caribbean Community in the 1980s.* p. 53.
52 *Ibid.*, p. 50.

W. Andrew Axline

6 Agricultural co-operation in CARICOM

The original condition of dependence of the countries of the Commonwealth Caribbean stems from their integration into the economy of the United Kingdom as the 'tropical farms of England'. Although in recent years this dependence has taken on other manifestations, the fundamental problem of Caribbean dependence and underdevelopment remains the agricultural sector. Through CARICOM the Commonwealth Caribbean countries are now mounting a regional response in the agricultural sector which contains elements of a production integration approach, even as CARICOM itself struggles to overcome crisis and stagnation in the wider sphere of regional co-operation. In taking this initiative CARICOM has moved agriculture from a position of low priority where it was seen as instrumental and secondary to achieving trade integration in manufactures to a position of high priority where development of the food sector has become a principal aim in itself.

The evolution of policy within CARICOM in respect of the agricultural sector is illustrative of the larger tendency of regional integration to move towards more *dirigiste* solutions to problems.[1] This evolution also reflects a shift towards greater concern for satisfaction of basic needs. The thrust of regional policy in the agricultural sector in the Commonwealth Caribbean has changed from one limited to individual sectoral agreements for specific commodities (rice, and oils and fats) to a comprehensive framework of integrated planning across the entire regional food and nutrition sector. The major elements of this policy include the Rice Agreement and the Oils and Fats Agreements, which antedated the establishment of CARIFTA in 1968; the Agriculture Marketing Protocol (AMP) and its modifications, including the Guaranteed Market Scheme (GMS); the Regional Food Plan, originally conceived as a major effort to achieve food self-sufficiency and now called the Regional Food and Nutrition Strategy, reflecting its broad concern with satisfying basic needs; as well as a number of financial programmes designed to help increase food production in the region. The role of these efforts in overcoming dependence will now be examined by looking at some of the major problems confronting the agricultural sector in the Caribbean and then assessing and evaluating the evolution of regional agricultural policy.

Problems and ostacles in Commonwealth Caribbean Agriculture

During the second world war, when shipping lanes were blocked, the region was able to go a long way towards feeding itself through efforts focused on satisfying local basic needs. Since then the problems facing the agricultural sector have worsened markedly. In the 1950s and 1960s the Caribbean received a large capital inflow which was directed mainly towards the industrial, extractive, and tourist sectors. In some cases this has resulted in competition with agriculture over land, but the larger impact has been on labour in the region. Higher wages in the extractive, industrial, and tourist sectors increased the reserve price of agricultural labour, without accompanying rises in agricultural productivity. The relatively low agricultural wages combined with the traditional disdain for agricultural work has produced high levels of unemployment side by side with labour shortages in the agricultural sector. As a result there has been a considerable population drift from rural to urban areas.

The effect has been to reduce considerably the area of land under cultivation. In Jamaica, for example, farming land declined from a total of 1,822,800 acres in 1958 to 1,489,000 in 1968, while 83,000 acres of land in farms over the size of 100 acres are also idle, of which 70% is land of the best quality. Throughout the region as a whole approximately half a million acres of cultivable land is lying idle or under-utilised.[2] It has consequently been estimated that only 32% of the population of the region is now supported by agriculture, with this figure dropping from 47·1% in 1953 to 33·7% in 1973 in Jamaica, from 37·1% to 20·5% in Guyana from 1960 to 1972, and remaining stable at about 20·5% in Trinidad.[3]

The productivity of agriculture is further limited by fragmentation resulting from division of holdings by inheritance and the geographical dispersal of a single farmer's land.[4] Of the region's 350,000 farm holdings, 95% are less than 25 acres in size, accounting for less than 30% of the total acreage in farms, while holdings of more than 100 acres account for more than 50% of the total acreage.[5] This situation where a small number of large estates hold a large proportion of cultivated land while a large number of small farms occupy a small share of the land results in a negligible amount of full-time family farming on medium-sized farms (25–100 acres). Owing to the general inadequacy of services small farmers also possess low levels of technical knowledge, which generates low output and high production costs.

Infrastructure limitations such as inadequate feeder roads and a poorly organised marketing system, which both emphasises the development of export agriculture and responds to the domestic political need for low consumer prices,[6] have further impeded the development of the local food sector. This sector suffers too from a lack of capital, resulting in dilapidation and neglect of farms, which has itself been aggravated by the shortage of adequate credit to small farmers and a continuing rise in the cost of agricultural inputs.[7]

Finally, a combination of other factors complete the panoply of problems facing agriculture in the Caribbean. They include adverse weather conditions and natural disasters, declining soil fertility, the advancing age of farmers, abandonment of land, lack of physical planning, the failure of agrarian reform, inadequate technical services and poor technology.[8]

Although agriculture remains the largest source of employment in the region, accounting for 10% of GDP and absorbing about 25% to 30% of the working population, it is manifestly in decline in the Commonwealth Caribbean. Agricultural stagnation pervades the entire region, even though the impact of this trend on the overall economy does vary from country to country, being most severe in the LDCs and Guyana.[9] *Per capita* food production has declined over the past fifteen years, with a resultant increase in the region's food trade deficit. Agricultural production for export fell by about 40% from the mid-1960s to the mid-1970s, reflecting the significant drop in the total amount of land under cultivation and the movement of population away from farming. From 1965 to 1974 the volume of regional sugar exports fell by more than 20% and banana exports by nearly 50%, with citrus also following the same pattern of decline.[10]

As a result of these trends the region has a large and growing deficit in foodstuffs which totalled US$450 million in 1978, or about US$80 *per capita*.⁹ The major food items composing these imports are animal proteins, meat, dairy and fish products, and animal feed, which make up more than 50% of the total value of food imports. Fruit and vegetables make up 9% and oils and fats 4%.[12] In the face of this growing food import bill member governments have initiated a variety of policies designed to increase self-sufficiency, diversify agricultural production and reduce the agricultural trade deficit,[13] but have not been able to prevent the problem reaching crisis proportions in the early 1980s. This is perhaps not surprising, given the magnitude and multitude of obstacles facing agricultural development in the region, but it also highlights the fact that until the last few years Commonwealth Caribbean governments have placed the main emphasis on industrialisation as the path to development. Moreover, those governments that did initiate policies of agricultural diversification and import substitution tended to pursue such rationalisation on a national scale, ignoring the regional context. Only since the inauguration of CARIFTA have serious attempts been made to overcome this deficiency of policy.

The early stages of regional agricultural policy: the Agricultural Marketing Protocol

In the initial phase of Commonwealth Caribbean integration regional agricultural policy reflected the minimalist *laissez-faire* approach characteristic of the beginnings of CARIFTA. When the free-trade agreement came into force on 1 May 1968, so did the Agricultural Marketing Protocol (AMP), which provided for protected regional trade in twenty-two selected regionally produced agricultural commodities.[14] The AMP constituted the extension and incorporation into a regional integration scheme of the principles which had been applied through sub-sectoral agreements for two commodities: rice, and oils and fats.

The Rice Agreements and the Oils and Fats Agreement had been in force since the second world war. In the case of rice, Guyana had three separate agreements: one with the Windward Islands, Leeward Islands and Barbados; one with Jamaica; and one with Trinidad and Tobago. Under these agreements Guyana would supply at an agreed price all the rice required by these territories, which in turn agreed to purchase only from Guyana unless that country was unable to satisfy their needs.[15] The Oils and Fats Agreement included all CARIFTA member territories except Jamaica and the Leeward Islands. It was substantially revised in 1970 and redrafted as a protocol to the CARIFTA Agreement, and its membership was extended to include all CARIFTA territories. The Oils and Fats Agreement provided for the fixing of area export prices of copra, raw oil and refined oil, the allocation of markets on the basis of declarations of surpluses and deficits of copra by member territories, and for the control of imports of oils and fats and substitutes from outside the area. The effect was to guarantee export prices and markets for the raw material and the semi-manufactured product.[16]

The sub-sectoral agreements in rice and oils and fats thus provided the background and experience for the adoption of the AMP as part of the CARIFTA agreement. As an integrative measure the AMP was seen as a means to create net benefits for the region within the agricultural sector, and as a means to redress the unequal gains from integration in the industrial sector, by creating benefits for the mainly agricultural LDCs.

As its name indicates, the AMP represented a market-oriented solution to problems in the agricultural sector. It designated a list of regionally produced foodstuffs which offer some scope for expansion of production and are now imported into the region in significant quantities, and in which some member territories have an export capacity. It aimed to regulate trade in these foodstuffs by allocating intra-regional production according to surpluses and deficits declared by the member countries, and by prohibiting extra-regional imports of these products and near substitutes when regional supplies were available. Market allocations were to be made for each commodity proportionately among member territories, based on estimated surpluses and deficits, and prices fixed at semi-annual meetings at a level designed to provide adequate returns to producers and to provide a reasonable price to consumers.[17] In short, the AMP was conceived to encourage agricultural development in the Commonwealth Caribbean as a whole by ensuring that commodities capable of being produced in the area are in fact produced and distributed, and to create special benefits for the LDCs of the region where small farming contributes a larger proportion to GDP.[18]

It is not surprising that the original policy instrument used to implement the goal of rationalisation of the agricultural sector was a mechanism to create a system of modified free trade. The AMP simply reflected the larger thrust of the Caribbean integration movement in the

Agricultural co-operation in CARICOM

direction of *laissez-faire* phased freeing of trade. The AMP was essentially a marketing agreement, and made no provision for regulating agricultural production in the region.[19]

It is now clear that the Protocol did not live up to expectations that it would stimulate the flow of intra-regional trade, particularly from the LDCs to the MDCs, although it was perceived to be a limited success up to 1972.[20] The implementation of the procedures of the AMP proved to be quite complex in practice, and served to accentuate some of the shortcomings of the policy, as well as to draw attention to conditions in the member countries which impeded its effective operation.

The promotion of intra-regional trade under the AMP depended upon knowledge of basic supply and demand conditions in the region so that efficient allocation could be made. This knowledge was simply not available in sufficient depth because of the absence of reliable data on production and consumption in the countries of the region. Moreover, where appropriate data were available, trade could be allocated only when the surpluses and deficits were officially declared to the Regional Secretariat and could be matched to demand. Often deficits were not declared, even though member countries were importing from third countries while declared surpluses existed in the region.

In addition to this, member countries were free under the provisions of the AMP to expand production in any of the commodities on the list in response to an increase in local demand and/or higher prices. Since the AMP price tended to be higher than the existing local price in many countries, especially the MDCs, and since the MDCs have a larger resource base on to which to expand production, the AMP has had the effect of stimulating production in the MDCs, thus thwarting both the goal of increasing regional trade and the goal of benefiting the LDCs.[21] Even though minor changes were made, such as giving LDC surpluses priority over MDC surpluses and improving the means of communicating surpluses and deficits, the system still continued to attract serious criticism from participating governments, especially those of the LDCs.

In response, in July 1972, the Council of Ministers set up the Guaranteed Market Scheme (GMS), under which the MDCs were required to purchase specific quantitites of specified commodities at AMP prices.[22] The MDCs also undertook to adopt whatever national measures were necessary to ensure that their capacity to absorb this production was not affected. The LDCs for their part undertook to produce the crops to meet their commitments, which was intended to ensure them annual increments in the amount purchased. In addition the MDCs agreed to provide technical assistance to help the LDCs meet their commitment under the agreement.

Although some successes were recorded in trade under the AMP, including carrots produced in St Vincent and tomatoes produced in Montserrat, there was little increase in total production or intra-regional trade.[23] A number of operational and organisational aspects of the AMP and GMS have been identified as contributing to the failure of the schemes to create the expected benefits, including the pricing system, the method of allocating surpluses and deficits, the role of marketing boards, payment difficulties, inadequate shipping facilities, poor quality control, and poor administration of the protocol.[24] More important, however, the experience of the AMP/GMS clearly demonstrated the incorrect implicit assumption underlying the scheme: that, given adequate market opportunities, increased agricultural production will be forthcoming. In the context of the overall *laissez-faire* thrust of the Caribbean integration movement it is not surprising that the agricultural programme was preoccupied with trade rather than production. Its failure starkly revealed that the limits to agricultural production in the region are based in problems more deeply rooted than the simple lack of an organised market. As the World Bank confirmed in its 1975 Regional Study of the Caribbean, 'The AMP as well as the GMS are likely to be more successful in encouraging output and exports of the LDCs if they were part of active efforts to deal with the structural, basic factors inhibiting production.'[25]

These structural problems pose great obstacles to agricultural development, and require more fundamental solutions than those contained in the AMP/GMS. An overwhelming emphasis has been put on the development of traditional export crops, to the detriment of small-farmer cultivation of foodstuffs. Domestic agriculture has been marked by fragmented holdings on poor land of declining fertility, resulting in low output and high production costs, a low level of technology, skills, capital input, research and credit facilities for non-plantation crops; an irrational wage and price structure in an inadequate marketing system; and a declining small-farm population of increasing age trying to survive in a society which lacks social facilities in rural areas and generally manifests an unfavourable attitude towards agricultural work.

The existence of fundamental structural obstacles to the development of local agriculture, in combination with inadequate national and regional policies for attacking these problems, led inevitably to a continued stagnation and decline in food production. By 1974 the Commonwealth Caribbean food deficit reached crisis proportions, with an annual food import bill for that year of US$500 million.[26] With major increases in energy costs and declining production in the traditional export sector, the deteriorating balance of payments of the region (with the exception of Trinidad and Tobago) underlined the need to embark on a major effort to restructure the domestic agricultural sector on a regional basis, and to do so through policies directed at the basic problems on the production side rather than being limited to market solutions.

Towards a policy for agricultural production

Within the regional integration movement in the Commonwealth Caribbean the measures currently being pursued to effect structural transformation of the agricultural sector range from *laissez-faire* policies to encourage greater output to the creation of a regional public corporation with the power to engage in production, trade, and a wide range of other activities designed to contribute directly to agricultural development. The latter is the centrepiece of the strategy, but was preceded by more intensive efforts to provide credit for agricultural production and to arrange for better and more technically accomplished use of agricultural inputs.

The first set of measures concerned the role of the Caribbean Development Bank (CDB). Since the inauguration of CARICOM, the CDB, which is now formally listed as an Associate Institution of the Caribbean Community, has come more and more to fulfil the role of a regional integration bank. The decision to create the bank was taken at the 1967 Georgetown meetings when the CARIFTA Agreement was negotiated, and it has always been regarded by the LDCs of the region as a major mechanism to benefit them. Data on the distribution of CDB financing confirm this perception. The LDCs of the region have received three-quarters (74·4%) of the 'soft loans' granted by the bank, and a total of 56·7% of the total net financing over the period 1970–79.[27]

The importance of agriculture among the priorities of the CDB is reflected in the fact that nearly a third (28·9%) of the total financing went to the agricultural sector, only slightly less than the amount devoted to infrastructure (30·8%), the sector receiving the largest proportion of the bank's funding.[28] These figures include equity investment and loans to private and public borrowers, both direct and indirect, and include export and domestic agriculture. The oft heard criticisms of the operation of the CDB, its stringent lending criteria and the long delays between approval and disbursement, were well founded in the case of financing small-farmer food production. Small farmers are among the potential recipients of CDB funds with the lowest net worth and are the least well equipped to meet the complex application procedures and to withstand the protracted delays in disbursement.

To respond better to the particular needs of local food producers in the region the CDB established the Farm Improvement Credit (FIC) scheme. This scheme, financed by a Canadian agricultural fund, provides loans of EC$3,000 to EC$100,000 on a medium to long-term basis at concessionary rates. The loans are aimed at upgrading agricultural production, and require collateral. All LDCs of CARICOM and the British Virgin Islands are participants in the FIC programme, which between 1970 and 1979 had disbursed a total of US$4·4 million out of US$7·8 million approved for twenty-six loans in the nine territories involved.[29]

However, the FIC scheme has not resulted in as large a number of projects or in as great an amount of financing as had been hoped, and eight of the approved loans, representing more than half the total financing, have gone to projects in Belize. From these data it is clear that small farmers in the eastern Caribbean have not been able to call upon the resources of the CDB to the degree that had been expected. The reasons for this failure are of concern to the CDB, and have been the object of study on the part of the bank. Even with the concessionary rates and small size of the FIC loans, small farmers of the eastern Caribbean often have not been able to meet the lending criteria, nor in many cases have they been able to initiate application procedures. It is also apparent that there have been major problems with the administration of the loans by recipient governments, including some difficulties with the collection of outstanding debts.[30]

Recognising the need to make access to agricultural credit easier for the small farmer, the CDB established the Agricultural Production Credit (APC) scheme in 1977. The APC scheme was aimed at overcoming the major deficiency of short-term production credit, which local governments have difficulty satisfying from their own resources. Using soft funds provided by the United States Agency for International Development (USAID), the APC programme provides short-term loans (up to eighteen months) ranging from US$200 to US$4,000 to small farmers in the LDCs. It was designed to permit the small farmer to upgrade production by purchasing seeds, fertiliser, pesticides and other agricultural inputs. Inadequate agricultural credit facilities are, of course, only one obstacle to the transformation of the local food production sector of the Commonwealth Caribbean economy. Although the CDB has become a major source of agricultural credit in the region, it can hardly yet be said to have had a major impact on the production of local foodstuffs by the small farmer.

A second set of measures undertaken recently with a view to confronting some of the structural problems of regional agriculture relates to the area of research and development. Historically, virtually all agricultural research carried out in the Commonwealth Caribbean was directed to increasing production in the export sector, whilst in the modern era the few scarce resources devoted to research in the region have been concentrated in the industrial sector, further contributing to the neglect of food production. It was not until 1975 that the Caribbean Agricultural Research and Development Institute (CARDI) was established as a regional institution to serve the research and development needs of the member countries of CARICOM. CARDI is the successor to the Regional Research Centre attached to the Imperial College of Tropical Agriculture since 1965, and is now a region-wide organisation responsible to the Standing Committee of the Ministers of Agriculture of CARICOM. The core budget of CARDI is funded by the member countries of CARICOM, with Jamaica and Trinidad and Tobago each paying one-third of the budget, Barbados and Guyana one-ninth each and the LDCs together providing one-ninth. Specific projects are also funded by various external donor agencies.

The programme of CARDI includes commodity research aimed at removing constraints on the production of various crops or livestock, agricultural training geared to upgrading levels of expertise and introducing new methods and techniques, and dissemination of information to meet the needs of agricultural technicians and planners in the region.[31] Although the efforts of CARDI cover the entire spectrum of agricultural production, including traditional export crops, increased local food production has been identified as the primary objective for the agricultural sector throughout the region, and CARDI's main thrust has been towards helping the small farmer by means of the improvement of farming systems.

The major programme which brings the regional agricultural effort to the level of the individual small farmer is CARDI's Small Farm Multiple Cropping Programme, a four-year renewable programme funded by a USAID grant of US$4·5 million. Under this scheme the field officers of CARDI carry out research on a small number of working farms – rather than at experimental stations – the results of which are intended to be integrated into a separate agricultural extension programme for the eastern Caribbean aimed at bringing the benefits of this new knowledge to all the small farmers of the region.

At present another sub-regional agricultural extension project is being carried out by a separate programme, the Caribbean Agricultural and Rural Development Advisory and Training Service (CARDATS), funded by the United Nations Development Programme and executed by the Food and Agriculture Organisation (FAO) of the UN. CARICOM is a co-operating agency of CARDATS and will eventually replace FAO as the executing agency. CARDATS is an integrated rural development programme providing direct services to small farmers relating to all aspects of farming without the preliminary research stage of CARDI's small-farm programme. In effect, it aims to create small farmers out of the plantation workers and marginal cultivators of the eastern Caribbean. It provides support and services in financing, marketing, supply of inputs and knowledge of farm management on a direct day-to-day basis, and hopes through this training experience to demonstrate that small farmers can be well-off and enjoy a relatively comfortable life, thereby attracting more people to the land and beginning to overcome the social repugnance felt towards agricultural work throughout the region.

Both CARDI and CARDATS represent regional efforts to bring about agricultural development through direct action at the individual level on the production side, reflecting an important advance from the *laissez-faire* regional market solution represented by the AMP. A further and deeper recognition of the need to pursue more direct policies to transform regional agriculture is found in the steps taken in the last few years to establish a programme of regional planning in the agricultural sector.

The Regional Food and Nutrition Strategy and the Caribbean Food Corporation

As noted earlier, by 1975 the regional food import bill had reached US$500 million. The starkness of this figure in a region whose economy was and still is predominantly agricultural provided a stimulus for more direct regional action to overcome the problem.[32] In a widely reported speech the late Prime Minister of Trinidad and Tobago, Eric Williams, drew attention to the need for region-wide efforts towards greater self-sufficiency,[33] and steps were initiated to begin development of a regional food plan. These efforts have developed into a Regional Food and Nutrition Strategy for the Caribbean, and the creation of the Caribbean Food Corporation (CFC), which has been described, albeit by officials of the integration movement, as 'the greatest co-operative venture in the history of the Commonwealth Caribbean'.[34]

The first steps were taken at a CARICOM Heads of Government Meeting in St Kitts, in December 1975, where a Working Party on Food Production submitted proposals for the establishment of a Regional Food Plan and Food Corporation with the general aims of maximising regional food production for local consumption by mobilising in the first instance unused and under-utilised agricultural and other resources. High priority was given to substantially increasing the allocation of resources to food production in the Windward and Leeward Islands.[35]

The adoption of the Regional Food Plan reflected a shift of approach towards the problems of development of Commonwealth Caribbean agriculture from an essentially *laissez-faire* market approach to a *dirigiste* approach directed at restructuring agricultural production on a regional level. The AMP/GMS, and the regional efforts to provide agricultural credit and research and development, were attempts to create the conditions for agricultural development, while the Regional Food Plan envisages direct action by regional institutions to bring about this development. It calls for a combining and sharing of resources across national boundaries, with a major role to be played by other regional institutions such as the CDB and CARDI, and specifically proposes the adoption of a phased project-by-project approach.[36]

Progress in implementing the Regional Food Plan has consisted of undertaking preliminary design studies to determine production deficits for a number of sub-sectors and production targets for particular member countries and the region as a whole; working out with each government a country programme, including pre-feasibility studies for governmental approval; preparing detailed feasibility studies once government approval is granted, in order to approach financial institutions (principally the CDB) for funding; and finally beginning the process of implementation once financial arrangements have been worked out.[37] A number of sub-sector projects have been identified for priority action, and are in various stages of progress: livestock including milk and other dairy products, poultry, beef, mutton and pork; fish and fish products; cereals and grain legumes; fruit and vegetables; spices and essential oils; agricultural inputs (including bulk purchasing); and oils and fats. Of these sub-sector projects few have yet reached implementation, although a number are in the study stage. Two large-scale grain projects, one in Guyana and one in Belize, have already harvested some crops, but both are undergoing modifications as a result of poor results.[38] Although progress on the sub-sector projects of the plan has not been as rapid as originally envisaged in 1975, there have been further significant developments in other aspects of the Regional Food Plan.

In 1976 the third Conference of Ministers responsible for Agriculture took a decision to request the CARICOM Secretary General to broaden the thrust of the Regional Food Plan to cover the area of rural development, with an emphasis on productivity and the redistribution of income, measures to improve the combination, quality and distribution of foodstuffs, and responsibility for health and other related activities.[39] In response to this request, which was endorsed by the Ministers Responsible for Health and for Education, the CARICOM Secretariat began to formulate a framework for a broad multi-sectoral food and nutrition policy for the Community as a whole. In the difficult circumstances which befell Commonwealth Caribbean regional integration in the mid-1970s this took some considerable time to develop, but eventually the so-called Inter-sectoral Committee, meeting in Barbados in October 1979 and August 1980, produced a draft multi-sectoral plan covering agriculture, health, education and communications, entitled *CARICOM Feeds Itself: a Regional Food and Nutrition Strategy. A Strategy for the '80s*.[40] This working document sets out the major objectives of regional programming in agriculture and outlines programmes and a process of consultation and decision-making by which these objectives may be attained.[41]

The document explicitly adopts the goal of providing for the basic needs of society through sectoral programming across a broad range of activities.[42] The proposed strategy is composed of nine Programme Areas scheduled to become effective over a period of ten years and to achieve the following goals:

1. Increased production and availability of food, especially nutritionally important commodities.
2. Increased regional food reserves.
3. Increased consumption of nutritionally important food by especially 'at risk' groups (children and expectant mothers).
4. Improved health status of mothers and children in the region.
5. Reduced incidence of nutrition-related diseases.
6. Development of more relevant and effective education at school and adult levels, especially in the areas of agriculture, food and nutrition, and health sciences.
7. Development of more trained technical personnel.
8. Development of active support and public participation in the programmes through a communications component.
9. Development of an effective organisational and management system for the Strategy.[43]

This basic needs approach to the satisfaction of food and nutrition requirements sets targets of providing from regional resources by 1990 a minimum of 80% of the energy and 60% of the protein consumed in the region, and making available on a daily basis in the regional market the minimum recommended *per capita* dietary allowance of protein and energy in the appropriate protein/energy ratio.[44] The central focus of the strategy is collective action to achieve greater self-sufficiency and self-reliance in food through the restructuring of regional supply to meet current and anticipated demand and the adjustment of regional consumption patterns to meet possible or potential supply.[45] This 'production integration' approach is spelled out in Programme I of the strategy, which focuses on increased food production and availability through a framework in which land resources are allocated to the production of nutritionally important food to meet current and future demand, giving consideration to the need to maximise foreign exchange savings, net farm gate income, and employment.[46] Five elements of this programme can be identified.[47]

The first is the increase in the production of agricultural commodities (including fish products) for the local and export markets, based on the provision of an effective agricultural extension service, of adequate inputs to farmers, including seeds, feed, fertiliser, insecticides and credit, of more effective incentive schemes, and of accelerated land reform programmes and the application of appropriate technology. The second element is the reduction of post-harvest food losses through action towards the development of better techniques for identifying these losses, of improvement of pest and disease control, of improvements in methods of harvesting, handling and pre-sale sorting, and of appropriate storage facilities. The third element is the creation of an adequate marketing infrastructure to bring about an increase in the quantity and an improvement in the quality of food available, through improvement of the capacity and efficiency of public marketing agencies, the establishment of effective market information systems, improvement in the food transport system, the provision of ancillary services such as market research, development and promotion; improvement of facilities available to small traders, and improvement in regulations affecting trade in agricultural commodities, including the AMP. The fourth element is an amelioration of the food processing infrastructure, with emphasis on the determination of appropriate processing technologies (multi-purpose and adapted to local resources), the training of skilled personnel, increased food fortification, and the establishment of physical facilities on a rational basis to increase processing capacity. The fifth element is the increased production of export crops to provide foreign exchange for production inputs and imports of non-local goods. Actions identified in this area include increasing the efficiency of production through appropriate technology, increasing the stability of the export market, and increasing the labour supply to the agricultural sector.

These five elements provide the broad outline of a proposed sectoral programme to increase food self-reliance and self-sufficiency within the larger Regional Food and Nutrition Strategy. The goals and means of the strategy are extremely ambitious, and their realisation will depend *inter alia* upon the creation of a complex organisational structure involving a series of regional and national institutions in the area of food, nutrition, health and education.[48] Many parts of the strategy are still in the formulatory stage, although implementation has begun in the sub-sectoral projects identified in the earlier Regional Food Plan, and the major implementing agency, the Caribbean Food Corporation, has been set up. The CFC provides an excellent example of the kinds of direct intervention and institutions that will be needed to bring about the restructuring of agricultural production.

The CFC was proposed at the same time the Working Party in Food Production submitted the Regional Food Plan to the Heads of Government Conference in 1975. The CFC was formally established in 1976 as an autonomous, regional, commercial organisation with an authorised share capital of US$40 million, subscribed by the member countries of CARICOM.[49] The broad objectives of the CFC are to mobilise, plan and implement all stages of agricultural production schemes in the region, to mobilise funds, technology and managerial skills from within and outside the region, and to organise and facilitate bulk purchasing of inputs, the marketing of agricultural outputs, and other services related to agricultural production.[50] The CFC is mandated to be active in the production, processing, storage, transportation and marketing of food. To carry out this mandate it is given the powers to make investments; establish, manage and operate enterprises; engage directly in the purchase, processing, transportation, marketing and distribution of products; engage in financial operations; and establish subsidiaries and enter into joint enterprises to achieve any of its objectives.[51]

In effect the CFC is intended to function as a holding company which will engage in a series of production and marketing operations, either individually or in joint enterprises with the private and public sector, and in co-operation with the CDB, CARDI, CARDATS and other regional agencies. The CFC is designed to be the principal implementing agency of the sub-sectoral projects in the Regional Food Plan by serving as a publicly owned but commercially oriented investment and development company. As such it aims to undertake activities at present carried out by relatively few companies in the private sector, most of them subsidiaries of multinationals, in a way that is compatible with the economic and social development goals of the member countries of CARICOM.[52] It is also envisaged that the CFC will attract and channel funding from donor agencies into regional projects aiming to restructure agricultural production.

The CFC is now in operation, although it was four years after its formal establishment in 1976 before sufficient qualified staff were assembled to enable it to set about implementing its mandate. Noting the fact that it had been 'slow to get off the ground', the group of experts who recently reviewed the future of Caribbean integration urged in their report that the CFC now be given 'the capability to carry out its functions'.[53] The matter is not, however, quite that simple. Several major issues relating to the CFC's ambitious objectives and powers require satisfactory resolution if the plan is to work.

The first is whether the CFC will adopt an 'investment' or an 'operating' approach. In its early years the predominant thrust will probably reflect the former, with the CFC taking a minority position in joint projects rather than taking a majority position and assuming responsibility for management. It is felt that for the moment the scarcity of management resources and the desire for private sector support favour such an approach.[54]

A second issue concerns the relationship between the public and private sector. Significant differences in the ideological orientations of Commonwealth Caribbean governments are reflected in their attitudes towards the role of private enterprise. It is officially hoped that the CFC can bridge the gap between business and government in the region, and enlist technical, managerial and financial support from the private sector. To do this, the CFC is committed to functioning on a 'commercial', that is to say, self-financing basis. Even so, the balance will not be easily struck: an indication of the underlying difficulty is found in the contrast between a leader of the business community in Barbados who denounced the CFC as unwarranted state intervention in the private sector and an editorial in the *Guyana Chronicle* stating that capitalist ideas must not be allowed to become entrenched in CFC projects.[55]

A third issue is the conflict between the commitment to operate on a commercial basis and the desire to take action to improve agricultural production in the LDCs. To put the point another way, regional projects in the agricultural sector must reconcile the need to employ under-utilised and idle resources in production for local needs, particularly in the LDCs, with the need to satisfy reasonable requirements of efficiency. This is a problem for all aspects of integration among developing countries, which must satisfy both the economic and the political requirement of policy-making. Previous experience within CARICOM of the fate of other corrective measures intended to help the LDCs would suggest that the conflict is more likely to be resolved in favour of the 'commercial' rather than the 'development' perspective.

A fourth issue is the degree of overlap and complementarity between the CFC and other regional institutions. The overall strategy requires effective co-operation between the CFC and the Caribbean Investment Corporation in the development of agro-industrial projects; between the CFC and the CDB in initiating studies, investing, lending and co-ordinating funds from donor agencies; between the CFC and the CARICOM Secretariat in responding to specific needs of the Regional Food and Nutrition Strategy; and between the CFC and CARDI in providing technical assistance for various projects. Ideally all these institutions should be mutually supporting; in reality the complex organisational structure proposed for the Food Plan is likely to expose the shortage of administrative skills.

A fifth issue derives from the relationship between the activities of the CFC and the national efforts of the member governments in the agricultural sector. The CFC is empowered to operate within any of the member territories, and its mandate to carry out regional projects is defined broadly to include projects in or involving more than one territory, projects involving one member territory and one regional agency, and projects in one territory whose products will be traded within the region. The creation of such national bodies as the Food and Agriculture Corporation in Trinidad and Tobago, with a mandate similar to that of the CFC, raises the possibility of awkward national – regional conflicts emerging. Considerable care will therefore have to be taken to assure co-ordination and complementarity between regional activities and those of the national governments.

These various issues must be resolved in practice as the activities of the CFC progress beyond the present preliminary stage. However, the fact that such problems exist is indicative of the considerable powers given to the CFC as a positive measure to bring about the structural transformation of regional agriculture. When considered in conjunction with the broad scope of the Regional Food and Nutrition Strategy, the extensive mandate of the implementing capacity of the Caribbean Food Corporation gives the Caribbean potentially the most far-reaching and ambitious programme for agricultural development to be found among integration groupings of developing countries.

Conclusion

In the overall process of regional integration CARICOM is at a point of stagnation, after having 'deepened' the integration process with the adoption of an external tariff and additional distributive measures in 1973, and after having failed to adopt a regime to control foreign investment and the transfer of technology in 1974. To date, CARICOM has successfully adopted a series of measures to create net gains for the region and measures to distribute the gains equitably, but has failed to satisfy the third vital requirement – adopting measures to reduce the negative effects of dependence. Within the Regional Food and Nutrition Strategy the implementation of sub-sectoral projects by the CFC at least provides the potential of achieving all three objectives within the agricultural sector. This evolution in CARICOM towards more intensive integration on a less extensive scale (limited to a single sector) is illustrative of the process of adapting regional integration to the needs and conditions of underdeveloped countries. Indeed, it has moved so far from traditional customs union theory that perhaps it is misleading to refer to it as integration. More correctly, it is a form of regional co-operation among developing countries (RCDC) that is sometimes referred to as collective self-reliance, a term that describes common action among developing countries to restructure their relationship with third countries, particularly the metropole, as part of efforts to bring about a more self-sufficient process of development.

If CARICOM succeeds in its efforts to move towards regional sectoral programming in agriculture despite the crisis in the overall integration movement it will be due partly to the fact that the efforts are limited to a single sector, and partly to the fact that agriculture has been less a focus of political controversy. The former point enables negotiations to be conducted around a limited number of specific projects and activities, which more easily permits compromises to be arranged around packages of measures likely to gain support from all participants, and also directs scarce regional administrative and technical resources towards the solution of a smaller number of problems within a single sector. The latter means paradoxically that the prospect of effecting meaningful agricultural integration in the Commonwealth Caribbean benefits from the emphasis that has historically been given to the industrial sector. For these two reasons there are grounds for hope that something substantive will develop out of the Food and Nutrition Strategy, although still the most fundamental question relating to regional agricultural policy is whether co-operation can proceed in this sector in spite of the continuing stalemate within CARICOM as a whole. Whatever the eventual outcome, the far-reaching proposals devised for agricultural co-operation in the Commonwealth Caribbean provide an example of one of the ways in which regional integration can be directed towards efforts to overcome dependence and underdevelopment in the Third World.

Notes

1 This argument is developed fully in W. Andrew Axline, *Caribbean Integration: The Politics of Regionalism*, London, 1979, pp. 32–63.
2 See E. R. St J. Cumberbatch, 'Prospects for Caribbean Agriculture', Paper prepared for the Second Caribbean Seminar on Scientific and

Technological Planning, UWI, St Augustine, Trinidad, 1976.
3   *Ibid.*, p. 9.
4   *Ibid.*, p. 7.
5   IBRD/IDA, *Caribbean Regional Study*, Report No. 566A, Vol. III, Agriculture, Washington, DC, 1975, p. 5. Caribbean Development Bank, *Small Farming in the Less Developed Countries of the Commonwealth Caribbean*, Barbados, 1980, pp. 319–21.
6   Curtis E. McIntosh and Michael Lim Choy, *The Performance of Selected Agricultural Marketing Agencies*, UWI, Department of Agricultural Economics and Farm Management, Occasional Series, No. 11, St Augustine, 1975, pp. 3–4.
7   K. A. Leslie, 'Contribution of Agriculture to Economic Development – a Case Study of the West Indies: 1950–1963', *Proceedings* of the First West Indies Agricultural Economics Conference, UWI, St Augustine, 1967, p. 12.
8   See Cumberbatch, 'Prospects for Caribbean Agriculture', p. iii.
9   For basic statistics on agriculture in the region see IBRD/IDA, *Caribbean Regional Study*, and United Nations, ECLA, Office for the Caribbean, *Economic Activity 1979 in Caribbean Countries*, CEPAL/CARIB 80/5, Port-of-Spain, 1980.
10 Sidney Chernick, *The Commonwealth Caribbean: The Integration Experience*, Baltimore and London, 1978, p. 119.
11 International Bank for Reconstruction and Development, *Regional Agricultural Development and Food Production in the Caribbean*, Report No. 2064-CRB, Washington, DC, 1978, p. 5.
12 Chernick, *The Commonwealth Caribbean*, p. 121.
13 Jamaica has pursued a number of policies aimed at stimulating agricultural production, including operation GROW (Growing and Reaping Our Wealth) in 1972. Trinidad and Tobago has created the Food and Agricultural Corporation, a state organisation designed to participate direct in agricultural production as part of an overall food strategy. Guyana has banned imports of foodstuffs for which substitutes can be grown locally, and experiments in farm co-operatives have been launched in Guyana, Jamaica, Montserrat and Grenada. The eastern Caribbean territories have been greatly disadvantaged in their ability to mount agricultural programmes because of the lack of research and administrative resources, and there has been virtually no agricultural planning on a national level in this sub-region.
14 They were carrots, peanuts, tomatoes, red kidney beans, black pepper, sweet pepper, garlic, onions, potatoes, sweet potatoes, string beans, cinnamon, cloves, cabbage, plantains, pork and pork products, poultry meat, eggs, okra, fresh oranges, pineapples, pigeon peas.
15 Prem Arjoon, 'Agricultural integration in CARIFTA: a unique experiment', *Guyana Graphic*, 13–14 September 1970.
16 Commonwealth Caribbean Regional Secretariat, *CARIFTA and the New Caribbean*, Georgetown, Guyana, 1971, pp. 33–4.
17 B. Persaud, 'The Agricultural Marketing Protocol of CARIFTA and the Economic Integration of Agriculture', *Proceedings* of the Fourth West

Indies Agricultural Economics Conference, UWI, St Augustine, 1969, p. 107.
18 Louis L. Smith, *Critical Evaluation of the Performance of the ECCM Countries under the Agricultural Marketing Protocol (AMP) and the Guaranteed Market System* (GMS), ECLA-POS 74/16, Port-of-Spain, p. 2.
19 George E. Buckmire, 'Rationalisation an Instrument for Development of Caribbean Agriculture', *Proceedings* of the Eighth West Indies Agricultural Economics Conference, UWI, St Augustine, 1973, p. 9.
20 Smith, *Critical Evaluation*, pp. 1–2.
21 *Ibid.*
22 *Ibid.*, p. 3. The particular commodities were peanuts, tomatoes, onions, cabbages, potatoes (not sweet), potatoes (sweet) and carrots.
23 *Ibid.*
24 *Ibid.*, p. 15. See also B. Persaud, 'The Agricultural Marketing Protocol', pp. 108–9; CARICOM/ECLA, 'Study of the Payments Arrangements for AMP Trade', ECLA/POS, Port-of-Spain, n.d., pp. 2–12; and Gloria Francis, *Food Crop Production in Barbados and its Response to CARIFTA/CARICOM and the AMP*, UWI, Institute of Social and Economic Research (Eastern Caribbean) Occasional Paper No. 2, Cave Hill, Barbados, 1975, pp. 36–46.
25 *IBRD Caribbean Regional Study*, Vol. III, 'Agriculture', p. 16.
26 Caribbean Development Bank, *Annual Report*, Bridgetown, 1976, p. 19–20.
27 Caribbean Development Bank, *Annual Report*, Bridgetown, 1979, p. 11.
28 *Ibid.*, p. 10.
29 *Ibid.*, p. 41.
30 Development Management Consultants, *An Evaluation of the Farm Improvement Credit Scheme*, Caribbean Development Bank, 1980. Only in St Lucia and in Belize does the administration of the FIC scheme seem to have been satisfactory.
31 CARDI Agricultural Information Services, *Caribbean Agricultural Research and Development Institute*, Bridgetown, 1980.
32 See Haydon Blades, 'The Regional Food Plan', *CARICOM Bulletin*, 1 August 1978, p. 20.
33 Eric Williams, 'The Caribbean Food Crisis', Address to the Caribbean Veterinary Association, Port-of-Spain, Trinidad, August 12, 1974.
34 Caribbean Community Secretariat, *CARICOM Feeds Itself: Basic Answers to the Questions Most Often Asked About the Regional Food Plan*, Georgetown, 1977, p. 23.
35 Haydon Blades, 'The Regional Food Plan', p. 21.
36 Caribbean Community Secretariat, *CARICOM Feeds Itself: Basic Answers*, pp. 8–10.
37 Haydon Blades, 'The Regional Food Plan', p. 23.
38 'The Regional Food Plan: a progress report', *CARICOM Perspective*, 2 May 1980, pp. 3–6.
39 *Ibid.*, p. 6.
40 CARICOM Secretariat, Inter-sectoral Committee, *CARICOM Feeds*

Itself. A Regional Food and Nutrition Strategy. A Strategy for the 80s, Working Document prepared for the Technical Group Meeting on the Regional Food and Nutrition Strategy, Kingston, Jamaica, November 24–28, 1980, Georgetown, 1980.
41 *Ibid.*, Preface, p. vii.
42 *Ibid.*, chapter 1, p. 7.
43 *Ibid.*, chapter 3.
44 *Ibid.*, chapter 3, p. 1.
45 Caribbean Food and Nutrition Institute, *The Caribbean Food Plan: An Economic Framework*, Mona, Jamaica, 1978, cited in *CARICOM Feeds Itself: A Regional Food and Nutrition Strategy*, chapter 3, p. 2.
46 *Ibid.*, chapter 3, p. 3.
47 *Ibid.*, chapter 3, pp. 3–9.
48 The elaborate proposed organisational structure is described in *ibid.*, chapter 11.
49 IBRD, *Regional Agricultural Development and Food Production in the Caribbean*, p. 9.
50 Caribbean Community Secretariat, *CARICOM Feeds Itself: Basic Answers*, p. 17.
51 *Agreement Establishing the Caribbean Food Corporation*, Article 3, Article 5.
52 Clyde C. Applewhite, *The Caribbean Food Corporation: Concept and Functions*, Barbados, 1979, pp. 3, 7.
53 The Group of Caribbean Experts, *The Caribbean Community in the 1980s*, Georgetown, 1981, p. 46.
54 Applewhite, *Caribbean Food Corporation*, pp. 13–14.
55 *CANA Wire Service*, 8 June 1977.

PART III

The international level

With independence comes an international personality and the opportunity to make foreign policy. For the Commonwealth Caribbean states this universally marked a consolidation of existing relationships rather than a new point of departure. Traditional ties with metropolitan powers have formed the core of emerging national interest and remain a pre-eminent concern, even when radical revision of them has been sought. That this should be so is not surprising, given the long history and the multiplicity of links between the Commonwealth Caribbean and the United Kingdom, the United States and Canada. Virtually no area of social activity is left untouched, though certain prominent interests can be identified, such as trade, investment, tourism, migration and aid. All these are of particular concern to Commonwealth Caribbean states in their search for economic development and confirm the priority allotted to this dimension in foreign policy formation in general. Foreign policy has thus been framed primarily to serve domestic ends rather than achieve specific international outcomes, though this has not been entirely absent, especially in the 1970s when Commonwealth Caribbean states, singly and collectively, sought a revision of the international system in the establishment of a New International Economic Order. However, this but serves to underline the point, practical engagement in this instance being to promote a more favourable international framework for domestic development rather than for any reason of abstract justice which has occasionally, and spuriously, been advanced.

Behind such a limited and normally low-profile perspective has, of course, been the question of size. No Commonwealth Caribbean state possesses in any measure the attributes generally considered necessary to sustain an active foreign policy over any length of time and none of them, excepting perhaps revolutionary Grenada, has shown any real inclination to mobilise its people on ideological issues in order to contest policies in the international arena. Guyana might be considered an exception in that its foreign policy has been engaged in a number of areas and its spokesmen have been active in a number of international forums. However, Guyana's circumstances, along with those of newly independent Belize, have been singular in the form of threats to its territorial integrity such as no other Commonwealth Caribbean state has had to contend with. It has also been dominated politically since independence by a chief executive who has sought to maintain power by mobilizing international opinion behind domestic policies framed to secure this end above all others. Such calculated manipulation is rare in the Commonwealth Caribbean though it does draw attention to the independent influence of personality on foreign policy formulation and execution, which can do much to explain, for example, Jamaica's relatively high level of international engagement under Michael Manley and Trinidad and Tobago's quiet diplomacy under the long tutelage of Eric Williams. It can also be used to account for more idiosyncratic behaviour such as that of Eric Gairy in Grenada: indeed, given that size itself determines the situation in the smaller islands where the chief executive simultaneously holds the key portfolios of finance, foreign affairs and defence, then similar behavioural aberrations in the future cannot be entirely ruled out.

Alongside the imperative of economic development and the constraint of size has been the overriding reality of geography as modified by the contests for power in the region. The Commonwealth Caribbean has had no option but to be engaged in this, though it has entered some areas reluctantly and has sought to preserve its separate identity on a range of issues. This is particularly apparent in respect of the immediate environment of the Caribbean region, where both CARICOM and the Caribbean Development Bank have been maintained as exclusive to the interests of the Commonwealth Caribbean despite requests to participate by Suriname, Haiti and the Dominican Republic, and where it was a collective decision of the Commonwealth Caribbean Heads of Government meeting in 1972 that diplomatic relations with Cuba be established. Towards non-contiguous Latin American states relations have developed very slowly, with meaningful content and reciprocity of interest being established only in the early 1970s. The leading edge here was provided by Venezuelan bilateral initiatives in the eastern Caribbean, followed at some distance in time by exploratory feelers from Mexico and Brazil. This has introduced a new dimension into regional affairs which has not yet been fully appreciated by many of the states concerned and least of all by the United States, which continues to maintain that its hegemony in the region must be both unqualified and absolute. To the extent that this may bring the United States into conflict with these powers, and given that country's past record of securing objectives unilaterally by armed intervention or destabilisation of governments deemed hostile, the tranquillity of the region is far from assured. And against such eventualities, of course, the Commonwealth Caribbean has very little defence other than an assertion of non-alignment as the basic foreign policy orientation and an espousal of the need for 'ideological pluralism' in regional affairs.

The combination of factors outlined above has defined four external environments which have been of particular significance for Commonwealth Caribbean states:

1. *Relations with the United States.* These have increased in importance and constitute at the moment of writing the critical environment. The reason is not only the underlying economic significance of relations but also the fact that the US has chosen to heighten its political and security profile in the region with the elaboration of the Caribbean Basin Initiative. For the Commonwealth Caribbean, as for other states in the region, this venture is likely to raise as many problems as it promises solutions.

2. *The area of traditional relations with the United Kingdom.* Since the 1960s these have steadily declined in importance for the more developed countries such as Jamaica, Trinidad and Tobago, Guyana and Barbados, though they still constitute the single most important relationship for the majority of the smaller islands of the eastern Caribbean and the remaining British dependencies. The direct relationship with the UK has also been much attenuated in the 1970s by its extension to the whole European Community, particularly on the key issues of trade and aid.

3. *The concentric circles of interaction which constitute the sum of relations with the region.* Working from the outside, these comprise relations with the Latin American mainland states, the non-Commonwealth Caribbean island and peripheral mainland states, and the Commonwealth Caribbean itself. This last inner circle constitutes a core by both the quality and the intensity of interaction within it, and must necessarily be analysed separately. It has, however, also become

an increasingly contested one as close bilateral relations are established by core states with those situated in other circles, notably Cuba and Venezuela. Whether this is of real advantage to the states concerned and how great a threat it poses to the integrity of the Commonwealth Caribbean core are among the most hotly debated questions of recent years.

4. *The global system*. This is the totality of relations among all actors in the international system. Within it the Caribbean occupies, largely as a result of the Cuban revolution, a special place at the centre of the lines which divide the world today – the East–West and the North–South divisions. As a consequence the Cold War and the development of the Third World are of particular significance. In respect of the former, the Commonwealth Caribbean has sought non-involvement; in respect of the latter, engagement and leadership, particularly within the United Nations and its affiliated agencies. To realise this it has stressed its heritage of African and Asian peoples as well as its geographical position in the Americas.

These four environments constitute the sum of the international relations of the Commonwealth Caribbean. Others may impinge – for example, Canada, the Commonwealth, and from time to time the socialist world. They are all, however, contingent on one or other of the above. And in their interaction with them, singly and collectively, can be found the specificity and the generality of what constitutes the foreign policy of each and every Commonwealth Caribbean state.

Ramesh F. Ramsaran

7 Issues in Commonwealth Caribbean–United States relations

Over the last two decades the countries of the Commonwealth Caribbean have become increasingly dependent on the United States. This has been the result of both economic strategy and general policy orientation – the latter dictated in some measure by geographical reality and in some cases by ideological preference. If the Caribbean as a whole is located within the 'back yard' of the United States, then it is equally true that the United States constitutes the 'front yard' for the small states of the Commonwealth Caribbean. Its influence on the political economy of this region is unavoidable and vast. This chapter assesses the nature and meaning of Commonwealth Caribbean–United States relations. It begins by setting out the extent of the region's dependence on the United States and the implications this has for national development in the area.

Commonwealth Caribbean dependence on the United States

Trade. Commonwealth Caribbean countries are critically dependent on the international economy. Trade (including tourism) forms a high proportion of gross domestic product. The relative size of their trade and the nature of the products on which they depend make them essentially price takers in the world economy. As a result of this situation Commonwealth Caribbean countries are easily buffeted by international commercial developments. Dependence on a narrow range of markets also tends to increase their vulnerability. The following section examines the emergence of the US as one of the region's major trading partners in recent years.

A glance at Table 7.1 shows that the US supplied 8% of Barbados's import requirements in 1955 as compared to 28% in 1978. The total

Table 7.1 *Selected Commonwealth Caribbean countries: share of total trade with the US, 1955–78 (%)*

Country	1955	1960	1970	1975	1978
Imports					
Barbados	7·6	13·1	21·0	19·0	27·9
Guyana	12·9	19·8	23·5	25·7[a]	22·5
Jamaica	20·5	24·2	43·6	37·4	31·8[b]
Trinidad and Tobago	9·5	13·8	16·2	21·6	20·6
Belize	33·9	39·7	33·2[c]	n.a.	36·0[b]
Domestic exports					
Barbados	2·2	3·1	21·8	36·3	32·9
Guyana	8·9	16·5	27·6	27·6[a]	20·1
Jamaica	15·9	26·2	53·3	38·3	45·2
Trinidad and Tobago	2·6	19·6	46·9	66·6	66·4
Belize	28·6	10·7	29·1[c]	n.a.	47·0

Notes.
(a) 1974. (b) 1979. (c) 1968.
n.a. not available.
Source. Official Trade Reports.

value of imports in this latter year amounted to B$175 million (19% of GDP). Of this figure, 20% related to food imports and amounted to roughly 25% of Barbados's total food bill. Machinery items contributed another 34%, while manufactured and other intermediate products accounted for 38%. Together imports from the US falling in SITC categories 5 to 8 (manufactured and semi-manufactured goods) amounted to 34% of total imports in these categories. On the export side, the share of total domestic exports taken by the US has also increased, though in recent years the percentage has tended to fluctuate from year to year. In 1978 the value was B$62 million (7% of GDP) or 33% of total domestic exports. Sugar and molasses contributed B$9·3 million[1] or 15% of the total. Manufactured goods accounted for the rest. In 1978 the US supplied 23% of Guyana's imports, as compared to 13% in 1955. Of the G$160 million (13% of GDP) worth of goods imported in 1978, the food component was 26%, capital goods 24% and other manufactured goods 33%. Domestic exports to the US in 1978 amounted to G$148 million (12% of GDP) or 20% of the total. Of this figure sugar and molasses contributed 22% and bauxite and alumina 74%. The value of Jamaica's imports from the US in 1979 was J$555 million (13% of GDP) or 32% of total imports. Food contributed

13% (or 32% of Jamaica's total food bill), raw materials 9%, petroleum products 5%, machinery 19% and other manufactured and intermediate products 52%. The US supplied 24% of total imports falling in SITC categories 5 to 8. The share of domestic exports going to the US increased from 16% in 1955 to 45% in 1979. The value in the latter year was J$645 million (15% of GDP). Of this figure bauxite and alumina accounted for 91%. Like the other territories Trinidad and Tobago has been buying an increasing share of its imports from the US, which supplied 21% of the total in 1978 as compared to 10% in 1955. Of the TT$979 million (10% of GDP) imported from the US in 1978, food accounted for 13%, raw materials 3%, machinery 44%, and other manufactured and intermediate goods 36%. Looked at in another way, we can say the US supplied 30% of the import food bill, and 36% of the manufactured goods requirements. Domestic exports to the US have increased dramatically since the 1950s—66% of the total in 1978, as compared to 3% in 1955. In the former year the value of TT$3,194 million was equivalent to 34% of GDP. Petroleum and related products accounted for 95% of the total, chemical products 3% and sugar 0·7%.

With respect to the position of the LDCs we do not have trade data for recent years which would permit us to make a clear statement. From the reports available, however, it is possible to make a few observations. There is no doubt that, because of its relative nearness and the wide variety of goods it produces, the US economy has become increasingly important as a source of supply, providing between 20% and 40% of import needs. In 1979 36% of Belize's imports originated in the US, which took 47% of its domestic exports. As a buyer of LDCs' exports the US appears to be less significant. The main export crops such as bananas and sugar are marketed under special arrangements in the European Community, while the CARICOM market absorbs a significant proportion of their agricultural production and assembled manufactures.

In recent years tourism has become an important foreign exchange earner for several countries of the region. Earnings from this industry plus remittances from nationals living abroad have played an important part in helping countries like Barbados, Jamaica and the LDCs to finance their widening trade deficits. Almost the entire economy of Antigua is centred around tourism. Barbados produces sugar and has been making great efforts to develop a manufacturing sector. In 1980, however, exports contributed only B$379 million to foreign earnings as compared to gross earnings from tourism of B$507 million. In Jamaica the gross contribution of tourism was US$242 million as compared to US$961 million for goods. As a source of tourists the importance of the US varies from country to country. As can be seen in Table 7.2 the US share of Jamaica's tourism declined from 81% in 1971 to 61% in 1980. The industry in the other countries appears to be characterised by more diversified sources. The relative importance of the US in Barbados's tourism has also decreased from the early 1970s, but in recent years still accounts for about 25% of stop-over visitors. US visitors to Antigua amount on average to over 40% of the total. In the case of St Kitts-Nevis the percentage is around 25% to 30%, while for Trinidad and Tobago it is over 30%. In the economy of the latter country, tourism is not an important industry.

Table 7.2 *United States residents as a percentage of total visitors to selected Commonwealth Caribbean countries*

Year	Jamaica	Trinidad and Tobago	Barbados	Antigua	St Kitts Nevis
1970	79·5	35·8	36·5	49·5	28·9
1971	81·4	32·7	36·2	45·9	26·5
1972	77·5	33·7	35·9	44·1	25·9
1973	77·8	33·4	33·7	43·3	26·5
1974	78·4	33·3	28·7	43·8	25·7
1975	75·1	31·6	24·8	42·7	25·3
1976	70·0	30·5	25·0	40·6	29·0
1977	67·9	30·6	26·1	42·9	n.a.
1978	66·4	31·2	27·0	n.a.	n.a.
1979	65·5	n.a.	24·6	n.a.	n.a.
1980	60·6	n.a.	n.a.	n.a.	n.a.

Note.
n.a. not available.
Source. CARICOM Secretariat, *CARICOM Statistics Yearbook*, 1978; official publications.

We have examined the importance of the US to the economies of Commonwealth Caribbean countries. But how important is the latter to the US in economic terms? As far as purchasing power is concerned these countries, even when taken together provide a very small market. In 1980 US exports to Commonwealth Caribbean countries amounted to less than 1% of US total exports, while imports from the group amounted to a little over 1%. The Caribbean, however, is a major source of bauxite for US manufacturers, though its importance has been diminishing in recent years. In 1979 Jamaica and Guyana accounted for 35% of the value of bauxite and alumina imported into the US, as compared to 66% in 1971. If we take the wider Caribbean, i.e. if we include the Dominican Republic, Haiti and Suriname, the proportion supplied by the Caribbean would be 48% in 1979 and 98% in 1971. These figures give an indication of the extent of the shift to sources such as Australia, which supplied 40% of the total in 1979.[2] Another strategic commodity traded is petroleum. The US relies on imported fuels to satisfy an increasing share of its domestic petroleum requirements (43% in 1979 as compared to 17% in 1960). Trinidad and Tobago exports petroleum and related products, but this trade is based largely on imported crude. Imports from this source amounted to about 2% of the total value of fuel imports into the US in 1979.[3] If we include the other Caribbean refining countries — the Bahamas and the Netherland Antilles — the proportion would rise to about 8%. The US, as indicated earlier, also buys a small proportion of the region's agricultural exports. The structure of protection in the US is such that manufactured and processed products encounter greater difficulty in penetrating this market. In 1979 it was estimated that 92% and 81% of US imports from Jamaica and Trinidad and Tobago respectively entered the US market duty free. Under the Generalised Scheme of Preferences (GSP) operated by the Americans some 87% of the exports from the Caribbean and the Central American Common Market countries enter the US duty-free.[4] These exports, however, cover only a limited range of products. Because of the complex procedures involved, the lack of capacity and the inability to satisfy origin rules, the Commonwealth Caribbean countries have not been able to take advantage of the GSP arrangements. Some of the products they are able to produce at competitive prices are excluded from the scheme or subject to quantitative restrictions.

With respect to tourism, we pointed earlier to its growing importance to the region as a foreign exchange earner. Tourism, however, is a volatile activity easily affected by economic conditions in the developed countries, transport costs, bad publicity, crime, political turmoil and the popularity of new centres. Some of these factors can be exploited with great effect by anyone with the resources and organisation to do so. In extreme cases, of course (as happened with the US and Cuba), a source country may prohibit its citizens from visiting a particular destination.

Because of the high import content and the degree of foreign involvement in Caribbean tourism, a great deal of the foreign exchange

Table 7.3 *US receipts from Caribbean and Central American visitors to the US (col. 2) and travel payments of US visitors to CAC countries (col. 3), 1970–79 (US$ million)*

			of which				
(1) Year	(2) US receipts[a]	(3) US payments[b]	(3) Bermuda	(4) Bahamas	(5) Jamaica	(6) Other BWI[c]	(7) Other WI and C. America
1970	170	390	63	127	95	44	61
1971	163	408	62	120	90	56	80
1972	169	504	69	144	105	60	126
1973	205	570	80	136	109	95	150
1974	216	685	110	151	122	87	215
1975	206	787	118	161	118	103	287
1976	289	784	133	168	109	125	249
1977	276	790	123	158	100	144	265
1978	332	888	136	198	118	153	283
1979	375	1,019	164	224	122	190	319

Notes.
(a) Excludes receipts of US carriers for transport of foreign visitors to and from the United States.
(b) Excludes payments by US residents to foreign carriers.
(c) British West Indies.
Source. US Department of Commerce, *Survey of Current Business*, various issues.

earned flows out again in the form of profits, payments for food and other materials (including souvenirs, building materials, etc.). The people of the Caribbean also add to the drain when they travel abroad. Table 7.3 gives an indication of expenditure in the US by visitors from the Caribbean and Central America (CAC) and receipts by the latter from US visitors in the period 1970–79. Expenditure by US visitors in CAC countries as a percentage of total expenditure abroad averaged about 11% in the period. On the other hand, US receipts from CAC visitors as a percentage of total travel receipts fluctuated between 4·4% and 7·5%. There are certain points we can make on the basis of this situation, and on observations made before. The fact that American tourists (like other tourists, of course) have a wide range of destinations from which to choose means that the Caribbean (despite its location advantage) does not have a monopoly of the 'tourism product'.

Competition places a limit on its ability to manipulate prices in order to increase income. Lack of control over an important cost component like air fares means that it (like other resorts) is to some extent a price taker. The Caribbean is also a buyer of the US tourism product, but a very insignificant one, and therefore as a bargaining ploy its value is of little consequence.

Private investment. The US is one of the world's major sources of private investment capital. US direct investment abroad increased from US$11·8 billion in 1950 to US$192·6 billion in 1979 at an average rate of 10·1%. In 1950 US$5·7 billion (49%) of these investments were located in developing countries, as compared to US$47·8 billion (25%) in 1979. There has also been a shift in industrial composition over the years. Petroleum's share dropped from 29% in 1950 to 22% in 1979, while the share of manufacturing increased from 32% to 43% over the same period.

In the post-war period North American transnational corporations have become involved in every sector of the Caribbean economies. They are to be found in banking, insurance, transport, construction, mining, manufacturing and even tourism. The amount of North American investment in each sector, or in individual countries, is not easily obtainable from published sources. Generally the figures tend to be subsumed in categories such as 'Latin America' or 'Latin America and Other Western Hemisphere'. Table 7.4 gives an indication of US private direct investment in various countries of what can be called the developing western hemisphere (DWH). The total US direct investment in DWH countries amounted to US$38·3 billion in 1980. Of this figure US$15·8 billion was located in South America, US$10·2 billion in Central America and US$12·3 billion in 'Other Western Hemisphere'. The latter included US$10·9 billion for Bermuda and US$962 million for Trinidad and Tobago, most of which is in the petroleum sector, which contains about 75% of total foreign assets. The estimate for Jamaica (in 1977) was US$556 million, US$221 million of which was in manufacturing. One writer gives the figure for Barbados as US$20 million and for Guyana US$27 million at the end of 1976.[5]

Given the present state of Caribbean economies, which we described earlier, it is clear that the post-war strategy of inducing foreign capital to effect economic transformation has had very limited success. The role of foreign capital in economic development is a very controversial one. A major criticism levelled against this policy is that in the long run the

Table 7.4 US direct investment position in Latin America and the Caribbean at end of 1980

Countries	Total	Mining	Petroleum	Manu-factures	Trade	Banking	Finance[a]	Other
1. South America	15,801	959	2,165	8,887	1,420	640	1,046	684
Argentina	2,446	(b)	399	1,548	213	129	16	(b)
Brazil	7,546	142	353	5,133	558	339	813	208
Chile	306	7	70	(b)	64	29	4	(b)
Colombia	961	(b)	217	547	97	12	23	(b)
Ecuador	321	0	158	114	32	(b)	−1	(b)
Peru	1,668	(b)	(b)	(b)	66	5	4	20
Venezuela	1,897	(c)	39	1,035	366	(b)	156	(b)
Other	655	1	(b)	307	23	59	31	(b)
2. Central America	10,163	138	868	5,157	1,386	210	1,548	856
Mexico	5,940	98	148	4,501	719	(b)	157	(b)
Panama	3,190	(c)	566	240	581	(b)	1,375	(b)
Other	1,033	39	154	417	85	(b)	17	(b)
3. Other western hemisphere	12,311	312	1,304	444	1,024	1,655	7,115	458
Bahamas	2,701	(b)	284	40	259	1,388	622	(b)
Bermuda	10,874	0	−203	15	(b)	(b)	10,313	(b)
Netherland Antilles	−4,072	(b)	(b)	(b)	16	(b)	−4,534	34
Trinidad and Tobago	962	0	(b)	(b)	7	(b)	3	47
Other	1,847	292	138	(b)	(b)	(b)	711	(b)
Total 1 + 2 + 3	38,275	1,408	4,336	14,489	3,830	2,505	9,709	1,998

Notes.
(a) Excluding banking, insurance and real estate.
(b) Suppressed to avoid disclosure of data on individual companies.
(c) Less than US$500,000.

outflows (in the form of dividends, interests, royalties, etc.) tend to exceed inflows. More and more capital inflows have to be induced to meet this drain, thus leading to greater dependency. In this connection it is worth looking at the Caribbean experience. In 1970, for instance, Barbados paid out B$13·2 million in investment income while net private capital inflows amounted to about B$24 million. In 1980, however, investment outflows totalled B$38·5 million while private capital inflows amounted to only B$28·2 million. Over the 1970–80 period net inflows of direct and portfolio investment amounted to B$283 million but the outflow of profits, dividends, interest, etc., was in the region of B$265 million. In the case of Guyana net private capital inflows in the 1969–79 period totalled G$62·3 million, while the outflow of investment income in the same period was G$527·6 million. This latter figure includes interest on the government's external debt, and therefore it is not possible to say what part is attributable to private companies or individuals. With respect to Jamaica, net direct investment inflow is estimated to have amounted to J$400 million in the period between 1970 and 1980. In the same period net direct investment income paid abroad was in the region of J$1,031 million. The experience of Trinidad and Tobago has been similar. In the period between 1970 and 1979 direct investment paid abroad totalled TT$6,559 million, while net direct investment inflows amounted to TT$2,673 million.

It is not possible to say what part of these inflows and outflows is associated with the United States. Data published by the latter, however, show that the income derived from its foreign investment tends to exceed annual capital investment. In Table 7.5 column 2 shows the income derived by branches of US corporations operating in 'Latin America and Other Western Hemisphere' countries. Column 3 shows the annual direct investment in these countries. It can be seen there that in the period between 1971 and 1980 annual investment exceeded income from past investment in only two years. In the period 1971–80 US branches invested some US$12·8 billion in 'Latin America and Other Western Hemisphere', but earned income totalling US$22·4 billion. It should be noted, however, that the figures in columns 2 and 3 do not include the reinvested earnings of incorporated affiliates of US enterprises. Such earnings, of course, would increase the foreign income derived by US concerns, but the capital outflow (from the US) figures would also increase correspondingly. Column 5 of Table 7·5 gives some indication of the size of these earnings.

Table 7.5 *Annual direct investment by US branches in 'Latin America and other western hemisphere' and the income derived from these investments, 1971–80 (US$ million)*

Year (1)	Branches — Foreign investment income (2)	Branches — Annual capital investment (3)	(3)–(2) (4)	Reinvested earnings of incorporated affiliates (5)
1971	1,130	691	−439	n.a.
1972	967	300	−667	n.a.
1973	1,622	673	−949	n.a.
1974	1,968	2,270	+302	n.a.
1975	1,603	1,347	−256	n.a.
1976	2,157	529	−1,628	1,323
1977	2,406	2,421	+15	1,582
1978	2,769	2,110	−659	2,097
1979	3,706	1,454	−2,252	2,589
1980	4,036	1,049	−2,987	3,229

Note.
n.a. not available.
Source. US Department of Commerce, *Survey of Current Business*, various issues.

Some observers argue that in order to appraise the role of foreign capital it is not sufficient to look only at capital inflows in relation to investment outflows in a particular period.[6] One has to examine the impact the original investments have had on the foreign exchange earning capacity of the economy as a whole. This observation has theoretical merit, but raises methodological issues which we cannot go into here. Even if we are prepared to accept it, it is clear that the attitude of foreign investors and domestic development policies would be critical factors in gaining maximum advantage from foreign resources. An examination of the Caribbean scene shows that despite a very accommodating policy towards foreign capital for most of its history, the economies remain highly undiversified, the main foreign exchange earners being tourism, bananas, sugar, oil and bauxite. Foreign investment has been largely of an enclave nature. Export agriculture and mining were in fact initiated and developed to serve essentially the needs of the metropole. Very little processing of the region's resources is done locally, and inter-sectoral linkages remain minimal even after many development plans. The manufacturing sector, as indicated earlier, is largely confined to the local assembly or packaging of products that were previously imported in finished form.

Aid. Over the years Commonwealth Caribbean countries have received a certain amount of financial, technical and material assistance from the United States under various programmes. Financial assistance, which takes the form of both loans and grants, flows through several channels. National institutions like the Agency for International Development (AID) and the Export-Import Bank have been the main channels for bilateral assistance. The US also provides funds through multilateral agencies like the World Bank Group, the International Monetary Fund, the Caribbean Development Bank, the Inter-American Development Bank, the Organisation of American States and the Pan-American Health Organisation. We do not have data on the quantum of this aid in a recent period which we can relate to overall capital inflows. Nor are we able to assess the significance of technical and other non-financial assistance made available. There are some figures, however, which permit us to give some indication of recent trends and increasing reliance on the US and on sources partly funded by the US. Table 7.6 shows that in the period between 1945 and 1980 three of the major Commonwealth Caribbean countries (Guyana, Jamaica and Trinidad and Tobago) received a total of US$469 million in US grants and credits, which is equivalent to about 0·2% of total grants and credits made available by the US in the period. At the end of 1966 Guyana's total external debt amounted to G$113·7 million, of which G$0·8 million was due to the World Bank, G$0·8 million to the US government and G$110·6 million to the UK government. At the end of 1980, however, the US component had grown to G$142·4 million out of a total external debt of G$953·9 million. The World Bank Group was due G$131·9 million, while the British government share was G$161·0 million. A similar trend can be observed with respect to Jamaica. At the end of 1970 21·3% of the outstanding national debt was repayable in sterling and 13·3% in US dollars.[7] At the end of 1980 the latter figure had increased to 27·7%, while sterling's share dropped to 3·7%. In the case of Trinidad and Tobago, out of a total external debt of TT$158·2 outstanding at the end of 1970, TT$67·1 million (42·4%) was due to the UK, TT$3·6 million (2·3%) to the Canadian government, TT$64·3 million (40·6%) to the World Bank, the US Export/Import Bank and the Inter-American Development Bank.

There were no loans outstanding to the Euro-dollar market. At the end of 1980, however, out of a total external debt of TT$1,047·9 million, TT$620·3 million (59·2%) was due to the Euro-dollar market, TT$2·7 million (0·3%) to the UK government, TT$18·7 million (1·8%) to the Canadian government, TT$114·3 million (19·9%) to the World Bank, the Inter-American Development Bank and the US Export/Import Bank. Miscellaneous loans accounted for the rest. At the end of 1973 Barbados had an outstanding external debt of B$55·4 million, of which only B$0·7 million was due to international institutions, and B$36·0 million to the Euro-dollar market. At the end of 1981 the total external debt had grown to B$259·4 million, of which B$59·6 million was owed to international institutions and B$30·7 million to the Euro-dollar market.

Table 7.6 US government foreign grants and credits, by countries, 1945–80 (US$ million)

Country	Post-war[a]	1945 –55	1956 –65	1966 –75	1976	1977	1978	1979	1980[b]
Guyana	102	(c)	7	71	7	6	4	3	3
Jamaica	201	16	2	119	8	7	19	19	12
Trinidad and Tobago	166	–	35	21	–4	–1	–2	29	89

(a) 1 July 1945 to 31 December 1980.
(b) Preliminary.
(c) Less than US$500,000.
– Zero.
(–) Negative figures occur when the total of grant returns, principal repayments and/or foreign currencies disbursed by the US government exceeds new grants and new credits utilised and/or acquisitions of foreign currencies through new sales of farm products.
Source. US *Statistical Abstract*, 1981.

The growing dependence on foreign funds is the result of a number of factors. One of the most important has been the inability to mobilise enough domestic resources to help finance development programmes. This is partly due to the misallocation of resources and the failure of past policies to have any significant impact on the productive capacity of the economy. It is also, no doubt, a reflection of the inadequate returns some of the countries have been able to obtain from the export of their natural resources. With respect to this latter factor it is clearly not easy for small countries acting alone (or even in concert) to negotiate equitable arrangements with transnational corporations, particularly, as we have said before, when these corporations have a considerable influence with their home governments and control important segments of world commodity transactions. In 1974, in order to increase its revenue from TNCs (mainly US) in the bauxite industry, the then government of Jamaica imposed a production levy and other conditions designed to keep production at a certain level. The Bauxite Production Levy Act had a marked impact on government's income from the bauxite industry, increasing revenue from J$24·5 million in 1973 to J$141·4 million in 1975. Subsequent events in Jamaica have shown that the ability of TNCs to curtail investment or to shift to other sources are powerful techniques in a situation where producing countries do not have strongly co-ordinated policies or an identity of interests.

The economic exchanges between Commonwealth Caribbean countries and the US have developed to the point where the former now find it necessary to peg their various currencies to the US dollar. In the context of present international exchange rate arrangements this decision has had a number of far-reaching ramifications. Firstly, it should be pointed out that pegging under a flexible exchange rate system is not the same as pegging under an arrangement in which all exchange rates are fixed by international agreement. The implication of daily changes in the exchange values of major currencies tends to be more complex. For instance, if the US dollar were to depreciate against the Japanese yen, CARICOM countries would find their import costs for Japanese products increased, since they would have to give up more local currency to acquire a yen. For export products quoted in US dollars the value of receipts (in terms of goods from non-US sources) would also have fallen. As far as debt servicing is concerned, a depreciation would result in lower servicing costs for US dollar-denominated debt, but higher costs for debt denominated in other currencies against which the depreciation had taken place. An appreciation, of course, would tend to have the reverse effects to a depreciation. The point we seek to make here is that since changes in the US exchange rate reflect developments in the US economy, and not in CARICOM countries, the latter would find themselves affected by, and reacting to, conditions which might have little relevance to their own domestic situation. Despite the growing ties with the US the search for economic independence must include the adoption of more appropriate exchange rate arrangements.

Emigration. North America is also an important destination for West Indian emigrants, particularly in the light of the restrictions imposed by the British government in recent years. Table 7.7 shows that between 1951 and 1979 over 360,000 people from the larger Commonwealth Caribbean countries emigrated to the US. In 1979 alone the figure was over 30,000, the large majority of them Jamaicans. Emigration is a double-edged sword. While the outflow of people may mitigate the effects which rapidly increasing populations put on public resources, and the remittances may help to alleviate balance of payments problems, the loss of skilled people (very often trained at public expense) is a severe handicap to the development of their respective native countries.

Table 7.7 US immigrants, by country of birth, 1951–79 ('000)

Country	1951 –60	1961 –70	1971 –79	1970	1975	1976	1977	1978	1979
Barbados	1.6	9.4	18.3	1.8	1.6	1.7	2.8	3.0	2.5
Guyana	1.0	7.1	39.1	1.8	3.2	3.3	5.7	7.6	7.0
Jamaica	8.7	71.0	123.1	15.0	11.1	9.0	11.5	19.3	19.7
Trinidad and Tobago	1.6	24.6	56.5	7.4	6.0	4.8	6.1	6.0	5.2

Source. US *Statistical Abstract*, 1981.

Trinidad and Tobago provides a by no means exceptional example of this problem. From 1962 to 1968 some 2,685 professionally qualified and 8,227 otherwise skilled and qualified persons emigrated to the US, Canada and the United Kingdom.[8] The costs to the economy were highlighted by the Prime Minister at the time.

> Between 1965 and 1969 we trained 696 nurses, but in the same period 586 nurses resigned and emigrated. The Trinidad and Tobago Electricity Commission operates one of the best training schemes in the country; in 1969 it trained 77 technicians and craftsmen of whom 33 emigrated. We have spent $4 million TT on scholarships of one sort or another for university training in the past five years, on the condition that each scholar agrees to serve the community for five years after his training; in the past five years 85 such scholars have defaulted on their obligations.[9]

In such a situation international technical assistance schemes can scarcely do more than stem the tide, certainly not turn it.

United States perceptions of the Commonwealth Caribbean

Even if the Commonwealth Caribbean had none of these ties of dependence with the United States, the latter could not be uninterested in what happens there. The Commonwealth Caribbean does have a limited economic importance to the US, although this element is only a small part of the larger geopolitical concern that guides American foreign policy in the region. Vaughan Lewis has suggested that US interests in the Commonwealth Caribbean consist of five separate but related concerns.[10]

Security. The United States considers the whole area of the Caribbean Basin (defined to include the rest of the non-Commonwealth Caribbean and Central America) as an important security zone, and views instability anywhere in the region as a threat to its own security. Although other areas of the world may be given a greater day-to-day significance in strategic terms, the Caribbean is deemed to represent the fundamental underpinning of the American security system. This perception colours the array of terms – the United States' back yard, the American Mediterranean or lake – used to describe the region in US government circles. The supply of arms to friendly governments and the deployment of American naval and military resources in the area follow naturally from this line of thinking. Moreover, since the revolutionary regime in Cuba is officially seen in the United States as the local proxy of the global communist system, a Cuban presence in any Caribbean country is perceived almost automatically as 'intervention' against US interests, and response made accordingly. This embraces not only Cuban military activity but also the provision of technical assistance and the establishment of diplomatic links with Caribbean governments. The active regional policy of the Cuban government in recent years has therefore contributed considerably to a raising of US security concerns with the Basin region, not excluding the Commonwealth Caribbean.

Communications. The Caribbean Sea constitutes an important communications route for trade to and from the United States. As President Reagan recently reminded the leaders of the Organisation of American States, 'nearly half of US trade, two-thirds of our imported oil, and over half of our imported strategic minerals pass through the Panama Canal or the Gulf of Mexico'.[11] The preservation of these lines of communication is obviously a high American priority.

Natural resources. The United States is especially concerned to ensure the uninterrupted continuation of trade in Caribbean mineral resources. Bauxite is at present the most valuable commodity for which the US has to rely on the Commonwealth Caribbean, although if one takes into account the view that oil and gas reserves in the region are far greater than is currently known, US interest in local minerals can be seen to be greater still. This interest is linked in American eyes to the question of local policy towards foreign investment and public ownership of strategic mineral industries.

Immigration. The United States is increasingly worried about the effects upon its own social, economic and political system of the movement – both legal and illegal – of Caribbean peoples to America. The majority are from Mexico, Central America and the Hispanic islands of the Caribbean, but the Commonwealth Caribbean is not excluded from this concern.[12]

Drugs. The United States is also alarmed by the illegal movement of drugs from and through the Caribbean states to its borders. This is a growing problem and one that Caribbean governments are reluctant to tackle because of the hidden balance of payments assistance such trade provides.

In a number of ways, therefore, US concern with the Caribbean mirrors the pattern of Commonwealth Caribbean dependence on the United States. Yet there are significant differences, notably the great American concern with security. Moreover, as Lewis points out, the last two areas of interest indicated above 'induce in turn a concern with the economic and institutional weaknesses in the structure of Caribbean political and social systems'.[13] There has thus grown up an American feeling that the governments of Caribbean states simply do not possess the capacity, either through smallness or weakness, to conduct themselves in a way amenable to the preservation of US interests. Traditionally this has led to a willingness to intervene in Caribbean affairs, either openly, as, for example, in the Dominican Republic in 1965 and in several Central American states over the last two decades, or subversively by 'destabilisation' of radical regimes after they have come to power, as occurred in Guyana between 1962 and 1964 and arguably in Jamaica between 1972 and 1980. Policies of economic assistance have been rarer, and where they have been introduced they have generally been characterised by a failure to comprehend the nature of the economic problem in question. There are certain legacies of Caribbean history such as concentrated income, land ownership and wealth, chronic unemployment, irrational production structures and alienation of large segments of the population which do not easily lend themselves to simplistic US-inspired economic solutions. The political culture associated with these problems has to be transformed to encompass the popular will and widen social goals. Even then vested interests do not yield easily. In these circumstances foreign assistance, as the Alliance for Progress showed, may benefit only privileged minorities.[14] Unbridled private enterprise, by making the rich richer and the poor poorer, has tended only to increase social tension and political instability.

The Caribbean Basin Initiative (CBI)

Against this background the programme of special assistance to the countries of the Caribbean and Central America announced by President Reagan on 24 February 1982 may be judged. The original impetus arose from a meeting of the US Secretary of State, Alexander Haig, and the Foreign Ministers of Canada, Mexico and Venezuela in the Bahamas in July 1981 to discuss the co-ordination of aid to the region. Officially heralded as evidence that the US government was sincere in its attitude towards Latin America and was not obsessed, as its critics alleged, with the security implications of underdevelopment, the plan soon ran into difficulties. Mexico made it clear that it was opposed to certain conditions the US was trying to include in the aid package, notably the exclusion of Cuba, Nicaragua and Grenada from its provisions. Canada, for its part, was also disturbed by the political overtones increasingly becoming apparent, and Venezuela steadily lost interest in the measure as a policy instrument, returning to its traditional bilateral dealings with countries in the region. When the CBI was finally unveiled, therefore, it was as a unilateral programme proposed by the United States and designed, Reagan assured Congress, to 'help revitalise the economies of this strategically critical region by attacking the underlying causes of economic stagnation' and making possible 'the achievement of a lasting political and social tranquillity based on freedom and justice'.[15] At the centre of such grandiose claims were the following measures:

The creation of a one-way free-trade area.[16] This was perhaps the most important element of the package. It proposed that exports (excluding textile and apparel products) should receive duty-free treatment in the US, even though the vast majority of Caribbean exports already enter the US market duty-free. The argument is that some of the duties that remain in force are in sectors of special interest to Caribbean Basin countries. They also limit export expansion into many non-traditional products. It was argued that the global reasons for excluding certain products from the US GSP system were not relevant to the Caribbean Basin. The complex structures of the GSP itself militate against the ability of small, inexperienced countries to take advantage of the opportunities offered. In addition, sugar would receive duty-free treatment but only up to a certain limit. For goods to qualify for duty-free entry they must have a minimum of 25% local value added. Inputs from all Basin countries could be cumulated to meet the 25% minimum.

Beneficiaries of the proposed free-trade area would be designated by the President. 'Communist' countries and countries which expropriate without compensation or which discriminate against US exports would not be eligible. The countries' attitude towards foreign investment and policies employed to promote their own development would also be taken into account.

Tax incentives. In order to encourage the flow of private capital to the area, the President proposed to ask Congress to provide 'significant tax incentives for investment in the Caribbean Basin'. He also indicated a readiness to negotiate bilateral investment treaties with interested Basin countries. The purpose of these treaties would be to provide 'an agreed legal framework for investment, by assuring certain minimum standards of treatment and by providing agreed means for resolving investment disputes'. Mention has also been made of the services provided by the Overseas Private Investment Corporation (OPIC), which currently offers political risk insurance for US investors abroad. This institution is in the process of expanding insurance coverage to eligible US investors by working with private-sector insurers to establish informal consortia to deal with projects on an individual basis.

Financial and military assistance. Non-military aid to Basin countries is currently channelled through three main programmes: (a) the Development Assistance Programme (DA), which is project-oriented; (b) the Economic Support Funds (ESF), which are more flexible and can provide direct balance of payments support as well as credit for crucial imports; and (c) food aid, provided through PL 480 programmes. For countries 'which are particularly hard hit economically' the President intended to provide additional funds in the fiscal years 1982 and 1983. It was proposed to increase the 1982 ESF current budget level from US$140 million to US$490 million. The proposed ESF figure for 1983 was US$326 million, while the DA figure was given as US$217·6 million, as compared to US$211·3 million budgeted for 1982. Total economic assistance (including food aid under the 'Food for Peace' programme) proposed for 1983 was in the region of US$664 million. It was stressed that the ESF funds 'would be used primarily to finance private sector imports, thus strengthening the balance of payments of key countries in the Basin while facilitating increased domestic production and employment'. Institutions like the IMF and the World Bank were to be consulted on the reforms necessary to ensure that ESF assistance had the desired impact.

Military assistance was treated separately. In the 1981 fiscal year the US provided military assistance of US$50·5 million to the countries of the Caribbean and Central America. El Salvador received US$35·5 million, or 70·3% of the total. The figure given for the fiscal year 1982 was US$112·1 million, of which El Salvador was to receive US$81·0 million, or 72·3% of the total. The supplemental appropriation of a further US$60 million proposed by President Reagan for 1982 would have brought the figure to US$182·1 million.[17] The 1983 estimate was US$106·25 million.[18]

Technical assistance and training. The CBI offered 'technical assistance and training to assist the private sector in the Basin countries to benefit from the opportunities of this programme'.[19] Efforts would be concentrated on investment promotion, export marketing, technology transfer, as well as programmes to facilitate adjustments to greater competition and production in agriculture and industry.

International assistance. Under the CBI, President Reagan still pledged 'to work closely' with Mexico, Canada and Venezuela 'to encourage stronger international efforts to co-ordinate our own development measures with their vital contribution and with those of other potential donors like Colombia'.

In addition to these and other measures proposed specifically for Puerto Rico and the US Virgin Islands, interest was expressed in helping Basin countries to modernise their agricultural and animal-producing sectors. It was hoped that the private sector would play a greater role in the development effort of Basin countries:

> The US government will be working with Caribbean Basin governments to design private sector development strategies which combine private, public and voluntary organisation resources in imaginative new programmes. We will also explore ways to promote assistance to comply with US health and sanitary regulations; to improve transportation links; and in general to remove public and private national and regional impediments to private sector development, with emphasis on new investment.

Reaction to and implications of the CBI

The programme received a mixed reaction in the region. The reason can be found as much in its context and timing as in its content. Until now the countries of the Caribbean Basin have been treated as part of the larger Latin American bloc. The decision to regard the Caribbean and Central America as a sub-region with special problems requiring special attention was a new approach which undoubtedly stemmed from recent developments in the area. The emergence of the socialist-posturing government of Michael Manley in Jamaica during 1972–80, the forcible removal in Grenada of the Eric Gairy government by the left-leaning Bishop regime in 1979, the violent overthrow of the right-wing dictatorship of Anastasio Somoza in Nicaragua, also in 1979, and the continuing political instability of El Salvador and other Central American states, were certainly major factors in the decision to formulate the CBI. The factors were perceived to be linked to deteriorating economic conditions in the region. It was reasoned that, if these conditions were improved, countries of the area would become more stable and thus less vulnerable 'to the enemies of freedom, national independence and peaceful development'. Whatever the rhetoric, then, the reality was an emphasis on the security dimension, with an explicit 'carrot and stick' approach built into the programme. Countries which pursued policies acceptable to the United States (e.g.

Jamaica under Edward Seaga) would be rewarded, and those (e.g. Cuba, Grenada, Nicaragua) which embraced views and programmes that did not accord with American perceptions would be denied assistance. The question of exclusion was handled in a very subtle way. The position was taken that the US did not explicitly exclude anyone from the CBI. Countries excluded themselves by virtue of the stance they adopted *vis-à-vis* United States values. 'We seek to exclude no one. Some, however, have turned from their American neighbours and their heritage. Let them return to the traditions and common values of this hemisphere and we will welcome them. The choice is theirs.'

The content of the programme itself drew a mixed response. Many of the politicians in the area welcomed the additional aid, though some of them (particularly in the eastern Caribbean) expected the financial part to be more substantial. Indeed, the amount of financial aid offered to the Commonwealth Caribbean within the initiative as a whole was not very high. Of the US$350 million supplemental aid proposed under the ESF programme for the 1982 fiscal year, Jamaica was to receive US$50 million, the eastern Caribbean US$10 million and Belize US$10 million. This was just under 20% of the total, as against almost 67% for three Central American countries (Costa Rica, Honduras and El Salvador). Political considerations seem to have weighed more heavily in this allocation than economic need, whilst the general disdain for development can be seen in the derisory total sum offered. In this connection it is worth pointing out that a technical group set up by CARICOM in 1981 estimated that Caribbean Basin countries would need some US$580 million in emergency aid alone in 1982 and US$4·7 billion in external financing over the five-year period 1982–86.[20] Figures of such magnitude have been conspicuous by their absence in all US discussions.

As far as military assistance is concerned, the total budgeted for 1982 amounted to about 20%[21] of all aid proposed for that year. (This does not include the additional US$60 million in military aid that was proposed by President Reagan for El Salvador.) The President stated that this expenditure was needed to meet 'the growing threat of Cuban and Soviet subversion in the Caribbean Basin'. There are many, of course, who would argue that the political instability of the region is rooted more in domestic political and economic conditions than in outside interference, and the situation is more likely to improve if these conditions are addressed direct. Failure to do this is likely to lead to annual escalations in military expenditure.

With respect to the trading arrangements, one view holds that the effect is likely to be more psychological than anything else, since 87% of Basin goods already enter the US market duty-free. Another view is that the non-tariff barriers would remain a serious impediment to an expansion of exports to the US market. A third position is that the duty-free market would be meaningful only if a country had the production capacity. The eastern Caribbean states, for example, would need to develop their physical infrastructure before they could significantly expand their output.[22] There may be some merit in each of these positions. Spokesmen for the Reagan administration, however, tend to see the effects of the free-trade arrangements in both a short-term and a long-term perspective. The immediate effects, they argue, would be felt in the traditional commodities (e.g. sugar, coffee, cocoa, vegetables, raw materials, etc.). This argument, however, has to be seen against the fact that in recent years earnings from most of these items have been declining, and not for lack of markets. In the medium and longer term, existing and new manufactured goods are likely to be affected. Again, it must be noted that many of the countries have not been able to satisfy the origin rule for manufactured goods to take advantage of the opportunities offered under various GSP schemes and under the Lomé Convention, in which several Basin countries are participants. The point is, the provision of markets may not be the crucial thing. Structural and technical problems exacerbated by irrelevant policies may be the more important factors inhibiting an expansion of production and exports. In the US itself the free-trade idea has encountered strong opposition in certain quarters, despite the safeguard provisions of the plan, and despite the fact that imports that would be affected by the proposals currently account for less than 0·5% of total US imports. The 25% local content requirement has been criticised as too low. Some American producers feel that the Caribbean will be used as a conduit by foreign competitors to penetrate the US market. The AFL-CIO group of trade unions is concerned about the impact on jobs if investors are attracted away by the proposed arrangements.

Finally, there is the impact the CBI might have on the Commonwealth Caribbean integration movement. The CBI shows no particular concern with integration objectives but rather addresses itself to an ideological drift and the need for the US to reassert its hegemony in the political and economic circumstances of the early 1980s. The result is that respected regional institutions such as the Caribbean Development Bank are likely to be downgraded and their aid policies politicised. There has already been a foretaste of what could happen. In 1981 the US sought to provide a US$4 million aid package to the Caribbean Development Bank to help the smaller Commonwealth Caribbean territories, on condition that Grenada was excluded from any benefits. The CDB refused to accept the funds on the grounds that such a condition was inconsistent with the terms of its charter, whereupon the Prime Minister of Dominica, Miss Eugenia Charles, promptly intervened to argue that Grenada should have excluded itself as a recipient country so as to allow the other states to benefit. In this instance she lost, but whether the lesson of unity has been learnt by all is far from clear. There is no common policy on foreign investment in the Commonwealth Caribbean, and in its absence member states of CARICOM could find themselves offering a wide variety of arrangements that would make nonsense of the whole integration movement. The industrial programming effort now being made could also be affected if foreign investors (with the collaboration of individual governments) decide to pay no attention to any agreements reached.

Conclusion

The benefits offered by the CBI are conditional. In other words, there is a cost involved, and prospective beneficiaries will have to decide whether they are prepared to pay it. More fundamentally, they will have to decide whether the benefits offered are significant for their development objectives, and whether the conditions are compatible with the solution to their economic difficulties as they see them. The positive aspects have to be weighed against the negative.

The basic strategy envisaged by the CBI is not new. It is cast in a particular framework which over the years has given rise to a great deal of controversy. A certain basic model is assumed in which government intervention in the economic system is played down, and a free enterprise system involving an expanded role for the private sector[23] is pushed to the centre stage of development strategy. When we add to this measures to attract foreign capital producing for a foreign market we have virtually all the elements of the Puerto Rican model. This explains both the support given to the CBI by business groups who tend to see the opening of the US market as an opportunity for expansion of trade and an increase in investment, and the opposition to it widely expressed by academics who note the consistent failure of the Puerto Rican model to deliver the goods. The ideological bias in the programme is also clear.

The fact that, to qualify for aid, domestic policies will have to pass the scrutiny of the US administration raises the whole question of political and economic sovereignty – a sensitive issue on which any advantages in the programme may eventually flounder.

As a result, it is possible to conclude only that the CBI does not remotely match up to the economic problems faced by either the Commonwealth Caribbean region in particular or the Basin as a whole. The initiative recognises that the political instability of the Caribbean has its roots in economic conditions, and it adopts economic rather than militaristic weapons to counter this. But in no way does it constitute a framework in which Commonwealth Caribbean dependence on the United States can be lessened. The reverse is the more likely and, from the US point of view, the intended consequence.

Notes

1. This B$9·3 million amounted to 17·4% of Barbados's total exports of rum and molasses in 1978.
2. This shift is taking place in the context of a growing dependence on imported bauxite and alumina which accounted for 93% of US domestic consumption in 1979 as compared to 74% in 1960.
3. Trinidad and Tobago's share of the US petroleum market has been falling. It was around 5% in the early 1970s.
4. See President Reagan's speech to the Organisation of American States on 24 February 1982, p. 4.
5. R. W. Palmer, *Caribbean Dependence on the United States Economy*, New York, 1979, p. 14.
6. See, for example, Isaiah Frank, *Foreign Enterprise in Developing Countries*, Baltimore, 1980, pp. 29–31.
7. It should be pointed out that not all debt repayable in US dollars is necessarily owed to the US government. Some may be due to regional and international financial institutions or to private capital markets.
8. Trinidad and Tobago, Central Statistical Office, *The Emigration of Professional, Supervisory, Middle Level and Skilled Manpower from Trinidad and Tobago 1962–1968 – Brain Drain*, Port-of-Spain, 1970.
9. Eric Williams, *Nationwide Broadcast to the People of Trinidad and Tobago, 23 March 1970*, Port-of-Spain, mimeo, p. 5.
10. Vaughan Lewis, *The United States in the Caribbean: The Dominant Power and the New States*, Presidential Address to the Caribbean Studies Association Annual Conference, St Thomas, US Virgin Islands, 27–30 May 1981, mimeo, pp. 6–7.
11. President Reagan's speech to the OAS, p. 3.
12. See G. D. Loescher and John Scanlan,' "Mass asylum" and US policy in the Caribbean', *The World Today*, 37, 1981, pp. 387–95.
13. Lewis, *The United States in the Caribbean*, p. 7.

14 See Jerome Levinson and Juan de Onis, *The Alliance that Lost its Way*, Chicago, 1970.
15 President Reagan's message to Congress on the Caribbean Basin Initiative, 17 March 1982, p. 2.
16 For a more extensive discussion of the CBI, on which this section draws, see R. Ramsaran, 'The US Caribbean Basin Initiative', *The World Today*, 38, 1982, pp. 430–6.
17 In a compromise move between the US House of Representatives (which baulked at the additional military aid to El Salvador) and the Senate, the additional appropriation was rejected during congressional discussions.
18 For the figures on financial and military assistance see State Department, Special Report No. 97, *Background on the Caribbean Basin Initiative*, March 1982.
19 Subsequent quotations relating to the content of the CBI are drawn from the above.
20 See communiqué issued by the Seventh Meeting of the Standing Committee of Foreign Ministers of the Caribbean Community, Belize City, 6 April 1982.
21 Compared to an actual total of 10·7% in 1981.
22 See the discussion of the CBI by various authors in *Foreign Policy*, 47, 1982, pp. 114–38.
23 As indicated earlier, most of the US$350 million supplemental assistance for 1982 was intended to finance imports for private sector development.

Paul Sutton

8 From neo-colonialism to neo-colonialism: Britain and the EEC in the Commonwealth Caribbean

On 28 February 1975 the first Convention of Lomé was signed between forty-six states of Africa, the Caribbean and the Pacific (ACP) on the one hand and the nine member states of the European Economic Community on the other. At the time the Convention was hailed as 'a breakthrough' in the way states of the industrialised 'North' would now relate to states of the developing 'South', and as a portent and example of what might be realised in the imminent debate on the 'New International Economic Order' (NIEO).[1] The four More Developed Countries (MDCs) of the Commonwealth Caribbean were very evident in the negotiations for the Convention, especially Sonny Ramphal of Guyana and P. J. Patterson of Jamaica, and clearly expected a great deal from it. Equally the Less Developed Countries (LDCs) of the Commonwealth Caribbean hoped to benefit and profit from the Convention through their varying degress of 'association' with Britain. The degree to which all countries of the Commonwealth Caribbean have done so is the main concern of this chapter. It constitutes a provisional assessment of the impact of Lomé I on the Commonwealth Caribbean, highlighting areas where there have been advances and, all too predictably, areas where there have been disputes. *Inter alia* it also assesses the Commonwealth Caribbean position in respect of negotiations for Lomé II.

The starting point: British membership of the EEC

The somewhat distant background to the Lomé Convention as far as the Commonwealth Caribbean is concerned was Britain's decision to renew its bid for membership of the EEC in the second half of the 1960s and the favourable response this received following the summit meeting of the 'Six' at The Hague in December 1969. Concern in the Commonwealth Caribbean over the likely effects of Britain's entry into the EEC turned into action and alarm as the real revision of the historical relationship with Britain, so long promised, appeared close to fruition.

What precisely this economic relationship was in respect of trade, and how entry to the EEC was likely to affect it, were set out at the time with admirable clarity by the London-based West India Committee in its publication *The Commonwealth Caribbean and the EEC*.[2] This pointed to real problems in respect of a number of commodities, notably sugar, bananas, rum and citrus, in which historical and special arrangements might be safeguarded only through the Commonwealth Caribbean individually seeking AOT status,[3] supplemented by some collectively negotiated guarantee for sugar. This course had clearly already recommended itself to some Commonwealth Caribbean countries, notably Jamaica, but not all, especially Trinidad. The latter's pattern of trade was somewhat more diversified than others', and it was felt that other forms of association, along the lines of the agreement negotiated between the EEC and the East African Common Market, might be explored as most appropriate in the long run. Both countries, plus the others of course, were at one on the question of sugar: here the visit by Geoffrey Rippon (the British Minister responsible for negotiations with the EEC) to the Caribbean in February 1971, and his later negotiated commitment from the EEC 'to safeguard the interests of the countries in question, whose economies depended to a considerable degree on the export of primary products, particularly sugar',[4] were critical. The Commonwealth Caribbean had made its point, and if it was not entirely happy about the 'good intent' of the EEC, it was able, through Robert Lightbourne of Jamaica, and in concert with others present at the Lancaster House meeting in June 1971, to underline its understanding of this 'as a firm assurance of a secure and continuing market in the enlarged community on fair terms for the quantities of sugar covered by the Commonwealth Sugar Agreement in respect of all its existing developing member countries'.[5]

On 22 January 1972 Britain signed the Treaty of Accession to the EEC. Under Protocol 22 the independent territories of the Commonwealth Caribbean were offered three forms of association: (a) association under a new 'convention of association' on the expiry of Yaoundé II (essentially AOT status); (b) some other form of association on the basis of Article 238 of the Treaty of Rome; or (c) a commercial agreement. The dependent territories of Britain were offered association under Part IV of the Treaty of Rome.[6] The question was, which form of association would the Commonwealth Caribbean collectively, or individually, take? Among the MDCs Jamaica and Trinidad already took opposing views; the LDCs by virtue of their 'association' with Britain were being offered a form of association not open to the MDCs; and further complications were now in the offing in the moves to 'deepen' the Commonwealth Caribbean's own integration process, which in itself raised a host of problems. In the end the Seventh Heads of Commonwealth Caribbean Governments Conference, meeting in Trinidad in October 1972, adopted a comprehensive negotiating position, leaving the matter open, but stressing the co-ordinated action of all Commonwealth Caribbean countries. The essence of this position was:

1. Caribbean Free Trade Area (CARIFTA) countries should seek a 'group' relationship with the EEC.
2. A *sui generis* relationship was required.
3. The focus of attention should be the content of the relationship rather than exercise of a particular option.
4. The relationship should provide for aid, at least for the LDCs.
5. CARIFTA should seek a secure and continuing market on the basis of an agreed Commonwealth quota for the supply of sugar to the enlarged EEC on fair terms in respect of quantities of sugar covered by the Commonwealth Sugar Agreement.
6. Commonwealth Caribbean governments must engage in intense diplomatic activity in Europe in order to convey the special problems of CARIFTA countries.
7. Invitations should be sent to representatives of the European Commission to visit CARIFTA countries early in 1973.
8. Closest co-operation should be developed with the African Associates (i.e. countries already associated under Yaoundé II) and other Commonwealth associables.
9. The position of the dependent territories should be clarified.

This position, especially as regards the first two points, was reaffirmed at the Eighth Heads of Government meeting in Guyana in April 1973. However, there was a new element in that extensive African contacts had raised the possibility of negotiating a fundamentally new 'association' with the EEC, not merely an enlarged Yaoundé II. The question now was whether this might best be realised through the

Caribbean negotiating as a group on its own (the argument of P. J. Patterson) or whether, while retaining its identity, it should join forces with Africa (the argument of Sonny Ramphal). In the end the latter position was adopted, with, not surprisingly, the choice of Sonny Ramphal to act as chief spokesman for the Caribbean in the impending negotiations with the EEC.

The negotiation of Lomé I: Caribbean concerns

When Ramphal rose to present the Commonwealth Caribbean's view at the opening of formal negotiations for the new convention in July 1973 he had behind him a considerable array of technocratic talent and political experience on which to draw. He also had the value of informal working relationships previously established with a number of African states. In mid-1972 a high-powered team of seven from a number of Caribbean countries, and including in particular Alister McIntyre, toured East and West Africa, gathering information and stimulating interest in how 'Commonwealth associables' might best relate to the EEC. One result, on the part of African countries, was the adoption by African heads of state of a common eight-point programme on negotiations with the EEC (the programme was apparently initially drafted by McIntyre).[7] Another was to set up a special secretariat to co-ordinate the African negotiating position and to invite the 'Commonwealth associables' elsewhere to liaise with this. According to one recognised authority on ACP/EEC relations, 'it was out of these contacts that the notion of an ACP entity rose'.[8]

What, in fact, is being emphasised is the very important informal role played by the Caribbean throughout the long negotiations for Lomé I. It has gone largely unrecorded, yet the extent to which African states had recourse to Caribbean expertise is a constant and frequent feature and, it might be surmised, an important aspect of the solidarity of the ACP as a whole. One more example may be cited to give the 'flavour' of this. In the period between July 1973 and the opening of real negotiations with the EEC in October, Ramphal agreed to step aside in favour of an African figurehead speaking with one voice for the whole of the ACP. This did not mean he was inactive, however, and to Ramphal must go the credit of drafting the reply to the statement of the EEC when negotiations opened in October – a reply ostensibly agreed to by the African states and actually delivered by their spokesman.[9] What other inputs the Caribbean provided 'off the record' in the 183 sessions of the

ACP and the EEC, the 493 meetings between the ACP themselves, and the 350 joint ACP-EEC documents, among much else,[10] is not known, but all Caribbean accounts suggest it was not inconsiderable.

As for the Caribbean itself, its position at the start of the negotiations was forthrightly set out in Ramphal's July statement.[11] This affirmed the Caribbean's intention to negotiate as a group[12] and its belief that an essentially new 'association convention' would need to be agreed. It also listed those areas where the Caribbean would particularly seek to influence outcomes, and these overwhelmingly, though not exclusively, focused on matters of trade. Given this, the Caribbean undertook a special responsibility to negotiate in this area. 'Aid' and Stabex, by contrast, were to be left largely to the African states.

The EEC's own 'semi-official' account of the Lomé negotiations notes that 'Title 1 of the Lomé Convention which deals with trade cooperation between the EEC and the ACP countries contains only 15 articles, most of which are quite short. This section, nevertheless, raised the toughest problems during the discussions and some of them were not resolved till the very final stage of the negotiation.'[13] Negotiations on sugar, bananas and rum likewise encountered difficulties, with the sugar protocol being the last to be agreed – after, in fact, the main text of the Lomé Convention itself had been adopted. The Commonwealth Caribbean, by design and circumstance, was thus in the thick of difficult negotiations, earning for its representatives a reputation as tough bargainers and a measure of respect from both the EEC and other ACP states for their diplomatic skills.

It is beyond the scope of this chapter to examine the details and intricacies of the negotiations. This is especially so in respect of sugar, which by all accounts was the most difficult to conclude and in which P. J. Patterson, as chairman of the sugar sub-committee, played such an influential part.[14] Overall, it is sufficient to note that in respect of the 'shopping list' presented by Ramphal at the beginning of negotiations, the Commonwealth Caribbean gained enough to be reasonably satisfied, collectively and individually. And it is this last aspect, in retrospect, which deserves some further comment.

At the beginning of negotiations the Commonwealth Caribbean as a group was beset by a political difficulty – the fact that some states were independent (and hence a party to negotiations in their own right) and others were not (the Associated states and other British dependencies). In respect of the latter group both the Commission and Britain maintained they could not be present during negotiations and further pointed out that they had no real need to be, since they were already guaranteed association under Part IV. The Associated states were far from happy with this but could make no headway in the matter. Moreover they operated under a thinly disguised British threat that 'if the Associated States should decide to reject Part IV they should bear in mind that what was not negotiable was a different arrangement for them alone'.[15] The Associated states of the Commonwealth Caribbean were thus obliged to fall back on the 'goodwill' of the MDCs. In the spirit of Caribbean integration then prevailing they were able, through the 'good offices' of Guyana and then through the direct intervention of William Demas, the CARICOM Secretary General, to have a representative of the East Caribbean Common Market (ECCM) sit with the official Caribbean delegation throughout negotiations.[16] In this way essential interests were safeguarded – particularly in respect of bananas.

Lomé I

The main features of Lomé I, which entered into force for a period of five years, may be summarised under four headings: commercial co-operation; financial and technical co-operation; industrial co-operation; and institutions. Below they are considered briefly in their general aspect and then at greater length in respect of their impact on the Commonwealth Caribbean.

Commercial co-operation. The Commonwealth Caribbean countries, as noted, were particularly concerned with the question of trade. Under Lomé I the following were the main principles agreed: free access to the EEC for almost all ACP products (over 99% of ACP products can penetrate the EEC with exemption from customs duties); non-reciprocity (the previous convention had insisted on reciprocity, which was resisted in particular by the Commonwealth Caribbean during negotiations for Lomé I and finally conceded by the EEC at the Ministerial conference in Kingston, Jamaica, in July 1974); softening of the rules of origin (of special significance to the Commonwealth Caribbean with its high-import-content industrialisation programme; special protocols on sugar, bananas and rum; and the devising of a scheme for stablising export earnings (Stabex) for twelve basic products. The effects of these on the Commonwealth Caribbean may be considered under the broad headings of trade, the protocols and Stabex.

Trade. Although the Commonwealth Caribbean has demonstrated a special interest in the trade relationship with the EEC, it should be noted that this is largely in respect of the region's dependence on the UK market for a few products critical to the maintenance of the present 'social order' in the region, viz. sugar and bananas. Outside this, the pattern of trade has for many years seen the declining significance of the UK and the growing significance of the USA as market and particularly as provider.[17] It has continued throughout the 1970s and has been especially marked among the MDCs, where the USA has continued to take a greater share of trade and the EEC in consequence a diminishing one.[18] Trade within the EEC is overwhelmingly dominated by the UK. This is evident by reference to the period 1975–80, where the UK was the major trade partner for all the MDCs (in the vast majority of cases accounting for over 50% of trade), the Ireland–Barbados sugar trade and the Trinidad–France–Italy petroleum products trade excepted.[19] (Incomplete data demonstrate this to be even more so in the case of the LDCs.) As expected, then, the overall pattern of trade has not changed significantly as a consequence of British entry into the EEC, confirming, indeed, that it was precisely to preserve existing patterns that the Commonwealth Caribbean sought association with the EEC in the first instance.

This is demonstrated even more precisely if the commodity structure of trade is examined. With reference to the MDCs the principal products exported to the EEC in the period 1976–79 were, for Barbados, sugar and clothing; Guyana, sugar, rum and bauxite; Jamaica, alumina, sugar and bananas; and Trinidad and Tobago, petroleum products and sugar.[20] Imports in each case were chiefly food products and manufactured goods (especially pharmaceuticals, chemicals and machinery).[21] In respect of several of the LDCs the case is as starkly drawn,[22] with Grenada exporting chiefly cocoa beans and bananas; St Lucia, bananas; and St Kitts-Nevis, sugar and molasses. As with the MDCs, manufactured goods and food products dominated imports.[23]

The commodity structure of trade revealed above is virtually identical to that uncovered by Adams in respect of the year 1973. Then he noted that twenty-nine commodities accounted for 'close to the entirety (about 97%) of exports from the Caribbean as a whole, and from the individual countries separately, to the EEC'. And 'that three commodities, viz. sugar (45%), bananas (17%) and bauxite and aluminium (15%) account for close to 80% of total Caribbean exports to the EEC'. The only other commodities of any note, raising the share to over 90%, were cocoa, rum, petroleum products, tar oil, spices and ammonia compounds.[24] This unchanging commodity concentration, in turn, naturally raises the question of the protocols, and in particular those dealing with sugar and bananas.

The protocols. The negotiation of separate protocols on sugar, bananas and rum attests to the particular complexities of ACP–EEC trade in these products. It is also reflected in the difficult negotiations surrounding each protocol and remains so in that in the case of both sugar and rum the outcomes were and are far from satisfactory to either side.

'If sugar did not exist, or if certain ACP countries did not produce it, perhaps there would have been no Lomé Convention.' This statement, accredited to an African diplomat,[25] spells out the importance of the protocol to Lomé as a whole. It is a 'keystone', and more time has been spent on considering sugar than any other single aspect of Lomé (the administration of financial and technical co-operation excepted). Certainly, today, its continuing significance is reported in the fact that the most dynamic sub-committee of the ACP Committee of Ambassadors is that pertaining to sugar.[26]

The sugar protocol has a number of features, the most important of which are a guarantee of purchase and price for an indefinite period. In respect of the first the protocol provides an undertaking by the EEC to purchase and for the ACP to supply each year agreed quantities of sugar. The total amount fixed for the Commonwealth Caribbean was 395,000 tonnes – a substantial reduction on the 696,000 tonnes guaranteed under the previous market of the Commonwealth Sugar Agreement, but one agreed to by the Commonwealth Caribbean at the time in that the general strategy was to maximise sugar earnings in a buoyant market by selling a greater proportion of sugar on the world market than before. The quantities apportioned to each Commonwealth Caribbean state were: Barbados, 49,300 tonnes; Guyana, 157,700; Jamaica, 118,300; Trinidad and Tobago, 69,000. Belize and St Kitts-Nevis were later to be given quotas of 39,400 and 14,800 tonnes respectively. The significance of this market for the sugar industry of the Commonwealth Caribbean is shown by the fact that in the period 1975–80 not one state defaulted in its supply (with the consequent penalty) – a situation not repeated in a majority of the other original signatories to the protocol.[27]

The vexed question for the Commonwealth Caribbean has thus been one of price. The protocol guarantees a minimum purchase price, negotiated annually, within the price range fixed for EEC sugar-beet producers. The Commonwealth Caribbean complains that the intervention price is not negotiated, but given, and that it is the lowest obtaining, determined without reference to the peculiarities of the ACP sugar trade. These complaints, in turn, rest on a background of falling world sugar prices and hence the failure of the price maximisation strategy noted above.

On the issue of negotiations the ACP undoubtedly have a case. With the exception of the first year of the protocol, when world sugar prices were still high, there has scarcely even been a pretence at negotiation.[28] The price is fixed within the Commission and by the Council, negotiating taking the form of informing the ACP as to its level, leaving them to accept or reject it. The spirit of 'consultation between equals' embodied in the Lomé Convention and the protocol has in this way been seriously compromised, and is in no way rescued by the argument strenuously urged by EEC officials, that this is all 'academic', i.e. the EEC price is a guarantee only, the reality being that most Commonwealth Caribbean sugar has been sold to Tate & Lyle at above the guaranteed price.[29]

On the question of whether the price is 'satisfactory', i.e. in some sense equivalent to the price deemed 'reasonably remunerative to efficient producers' of the CSA, no answer can be given here, owing to the complexity of calculation around which controversy abounds. Again the ACP sugar producers have a strong case in arguing that due allowance should be made for freight costs, and certainly the lowest intervention price has consistently been applied to them. Against this must be weighed the collapse in the world market price of sugar and the advantage now accruing to the ACP of a guaranteed market in the EEC. Here a calculation can be made for the Commonwealth Caribbean from figures set out in Table 8.1. They show that for each year of the protocol's operation, except 1975–76, the Commonwealth Caribbean has gained substantially from this arrangement to the collective figure of some 137,861,274 ECU.[30] Whether this should rightly be considered as 'aid' (the position of the Commission) or as 'chance' (the position of the ACP) is, of course, a matter of dispute.

Finally, it must be noted that the protocol was concluded for an indefinite period though containing provisions for renunciation and renegotiation. The latter has so far not been attempted. This bears witness to the 'value' of the protocol to the ACP, problems notwithstanding, and to the undoubted difficulties surrounding any future negotiation. These (if past and present are any guide) are likely to involve political far more than purely economic questions.

Moving on to the banana protocol, this has three fundamental aspects. The most immediately important was the statement that 'as regards its exports of bananas to the EEC, no ACP state will be placed, as regards access to the markets and market advantages, in a less favourable situation than in the past or present'. This was buttressed by (a) a promise of joint endeavours by the ACP and the EEC to devise and implement appropriate measures to increase banana exports to traditional EEC markets, and (b) a promise of comparable endeavours to enable ACP states to gain a foothold in new EEC markets and to extend their banana exports to them.[31] The protocol therefore embodied both protective and expansive features at one and the same time.

Table 8.1 *Sugar Protocol and world sugar prices, July 1975–June 1980*

Year July–June	(1) Guaranteed price (ECU/tm)	(2) Mean spot price, London ($US/tm)	(3) Mean exchange rate (ECU–$US)	(4) Spot price London (ECU/tm)	Difference, (1)–(2)
1975–76	255,300	342,500	1,159	295,513	−40,213
1976–77	267,300	219,900	1,114	197,397	+69,603
1977–78	272,300	189,000	1,192	158,557	+113,943
1978–79	278,100	199,500	1,327	150,339	+127,761
1979–80	341,300	426,100	1,397	305,011	+36,289

Hypothesis
1975/76 −40,213 ECU × 448,500 tm = −18,035,530 ECU
1976/77 +69,603 ECU × 448,500 tm = +31,216,945 ECU
1977/78 +113,943 ECU × 448,500 tm = +51,103,435 ECU
1978/79 +127,761 ECU × 448,500 tm = +57,300,808 ECU
1979/80 +36,289 ECU × 448,500 tm = +16,275,616 ECU

Source. Commission of the European Community, *La CEE et la région des Caraibes Lomé I et Lomé II*, Brussels, 15 February 1982.

The protective features, in fact, did no more than ratify existing national regimes, all of which had evolved in different directions in accordance with varying historical and economic circumstances.[32] In Britain the pattern that had thus emerged was one of Commonwealth preference plus quantitative restrictions on imports from dollar sources.

This gave a particular advantage to supplies from Jamaica and the Windward Islands, which on occasion in the late 1960s were supplying up to 97% of the British market.[33] Failure of the Commonwealth Caribbean to maintain this level in the mid-1970s has been a constant preoccupation of the countries concerned, where the highest market share was in 1978 at an estimated 63%. A number of reasons may be cited for this, but one of them is not market penetration by the other ACPs, this being constant at around 20%.[34] In fact banana markets in the EEC appear remarkably stable as to quantity, pointing to the fact that they are in essence 'managed markets' in which a few transnational companies predominate. The pattern for Britain is set by Fyffes (United Brands) and Geest, which between them consistently take around 75% of the market.[35] Developing new markets, or expanding existing ones, in this situation is clearly no easy matter. It is also predicated upon an expanding production in the producer countries, which in the Commonwealth Caribbean has not materialised.

The decline in production in Jamaica and the Windward Islands has been marked, with both countries exporting by volume some 50% less in 1973 than they had in 1965. While this is not without significance in Jamaica it is greatly magnified in the case of the Windward Islands. There bananas in 1974 accounted for over 50% by value of the total exports of Dominica, St Lucia and St Vincent. The industry is vital to the islands, and by common consent protection of the British market and expansion of production are absolutely crucial. This accounts for the identity of interest displayed by the UK and the Commonwealth Caribbean in the Chris International Foods case[36] and the 'enthusiastic' British funding of the Five Year Banana Development Programme in the Windwards. On the basis of a British aid disbursement of up to £4·3 million the plan is to raise production to 165,000 tons and to improve quality.[37] This level of British funding once again underlines the fact that the banana protocol is essentially a bilateral affair – a situation which allows all parties to be relatively satisfied with the protocol's working.

As with the other two protocols the rum protocol seeks to preserve the traditional pattern of trade. It does so by fixing an annual overall quota determined on the basis of the largest annual quantities imported by the EEC from the ACP during the last three years for which statistics are available. These quantities may be increased at an annual growth rate of 13% for the other EEC members and 40% for Britain.[38]

The British quota is by far the highest (around 72%) and is filled by the Commonwealth Caribbean in the following order of decreasing importance (1976–79): the Bahamas, Guyana, Jamaica, Trinidad and Tobago, and Barbados.[39] The fact that the Bahamas do not produce sugar cane but import molasses to make rum, which, in the absence of a common EEC definition of rum, they can then export with advantage to the EEC, is but one area of dispute between the Commission and the ACP. Another is the method of calculating quotas, with the EEC basing its figures on rum released from bond and the ACP on amounts actually exported to the EEC. All this, it must be noted, takes place against a background of ACP states failing to meet quotas and a belief by the Commission that the alcohol market in the EEC is already 'saturated'. Clearly, the rum protocol does not operate smoothly and is unlikely to do so in the absence of a common EEC market in alcohol.[40]

Stabex. The stabilisation of export earnings (Stabex) is generally considered the most innovative feature of the Lomé Convention. The aim of Stabex is defined as 'to provide a remedy for the adverse effects of unstable export receipts and thus to help the ACP countries to secure economic stability, profitability and steady growth'.[41] It works not by stabilising the prices of a range of products exported to the EEC but by guaranteeing a stable level of earnings determined according to specific criteria. In the first Lomé Convention 375 million EUA was earmarked for Stabex (plus 20 million EUA for the Overseas Countries and Territories (OCT)[42]) to cover twelve products. Only bananas and timber in the original list of products were of any real interest to any Commonwealth Caribbean country, which brings into focus a cardinal fact about Stabex — that it was specifically designed to be of particular benefit to African states, thereby representing a trade-off against the protocols for Caribbean states. This linkage of Stabex and the protocols is not officially acknowledged, but is there nevertheless, not least in the mention of both under Title II of the Lomé Convention.

Given this political background it will be no surprise to find that the Commonwealth Caribbean has benefited from Stabex only to a small degree. From 1975 to 1979, of some 389,811,700 ECU made available, only 3,235,300 ECU (0·82% of the total) went to the Commonwealth Caribbean. This represented payments of 342,400 ECU to Belize in respect of sawn timber and 2,892,900 ECU to Dominica in respect of bananas.[43] Hurricane damage in 1980 saw an increase in the amount disbursed, with 1,800,000 ECU going to Dominica, 2,200,000 ECU to Jamaica, 1,300,000 ECU to St Lucia and 700,000 ECU to St Vincent, all in respect of bananas.[44] As LDCs Dominica, St Vincent and St

Lucia will not be obliged to repay these amounts.

Financial and technical co-operation. As noted earlier, the Commonwealth Caribbean left the negotiations of 'aid' for Lomé I largely to the African countries. Their interventions in this area were thus limited, mainly technical, and especially concerned with increasing the role of the ACP states in the administration and management of 'aid'.[45] In this they were not as successful as they wished, and the same may be said of the 'aid' negotiations in general. By general agreement these were left until last. This allowed the EEC to present their figure for 'aid' very much as a *fait accompli* which the ACP could do little but accept even though it was less than half the sum they had proposed at the Kingston Ministerial conference. The final agreed figure of 3,550 million EUA was made up of 3,000 million EUA to the ACP and 150 million EUA to the OCT to be disbursed by the Fourth European Development Fund (EDF); and 390 million EUA to the ACP to be disbursed by the European Investment Bank (EIB). Of the total the UK was to contribute 18·7%.[46]

The total amount of 'aid' allocated to the Commonwealth Caribbean is given in Table 8.2. This, as can be seen, is divided under the two main headings of the Fourth European Development Fund and 'Other Interventions'. It will be so considered below, but before proceeding one fact should be borne in mind – the total EDF allocation of 93·2 million EUA is only a fraction (4·5%) of the total programmable 'aid' available under this head in Lomé I.[47]

The Fourth EDF. The basis of national allocations is the indicative programmes drawn up between the EEC and each ACP state. In the case of the MDCs a programming mission visited the region in September and October 1975 and a number of agreements were concluded. The amount of funds made available to each state is determined within the Commission according to general criteria (principally a combination of population and GNP), subject to a political review at the highest level, where the final figure is also decided.[48] The figures given for the MDCs of the Commonwealth Caribbean suggest a broad conformity to this arrangement, indicating that within the Commonwealth Caribbean no special political problems were then seen to exist. Of the total 47,499,000 ECU made available to them the individual share out was broadly proportionate, as follows: Bahamas, 1,800,000 ECU; Barbados, 2,600,000; Guyana, 12,800,000; Jamaica, 19,999,000; Trinidad and Tobago, 10,300,000.[49]

Table 8.2 *Commonwealth Caribbean: total EEC aid—Lome 1, 1975–81*

	ECU ('000)
1. *Financial and technical co-operation*	
MDCs,[a] indicative programmes	47,499
LDCs,[b] indicative programmes	19,660
Regional programme	26,000
Total Fourth EDF indicative programmes	93,159
2. *Other interventions*	
Exceptional aid[c]	8,425
Food aid	20,356
Non-governmental organisations	1,874
European Investment Bank	24,950
Special action (Guyana)	2,167
Total other interventions	57,772
Total EEC intervention (1+2)	150,931

Notes.
(*a*) Bahamas, Barbados, Guyana, Jamaica, Trinidad and Tobago.
(*b*) Dominica, Grenada, St Lucia, St Vincent, Anguilla, Antigua, Belize, Cayman Islands, Montserrat, St Kitts–Nevis, Turks and Caicos, Virgin Islands.
(*c*) 1976–September 1981
Source. Commission of the European Community, *La CEE et la région des Caraibes Lomé I et Lomé II*, Brussels, 15 February 1982.

As to the content of the programmes themselves, they are agreed by negotiations between the EEC and the ACP state, based on the initiative and recommendations of the latter. The breakdown by sector of this suggests both distinctive interests and a commonality of interest in operation. That is, whilst all the MDCs have emphasised as their first priority the development of production, second priorities have ranged over the provision of economic infrastructure, social development or trade promotion.[50] Finally, note should be taken of the relatively high allocation of aid via commitments (over 90%) except for the case of Trinidad and Tobago (55%). One authority has argued altruistic reasons for this, i.e. Nigeria and Trinidad and Tobago have declared their willingness to forgo EDF grants in favour of least developed countries.[51] While not discounting this, research in Port-of-Spain, however, indicates 'bureaucratic inertia' to be at least as significant a factor.

With the exception of Grenada, all the LDCs at the time the Lomé Convention was signed were considered as dependencies of Britain and therefore obliged to seek funds from the amounts made available to the OCT. Here they were placed at an immediate disadvantage by the formula on which the OCT funds were to be shared. This was to be equally between the British, French and Dutch dependencies wherever they existed and without regard to population numbers or levels of development. The Netherlands, for example, thus managed to secure for the Netherlands Antilles (population 248,000 and in 1978 in receipt of US$195 *per capita* foreign aid) an 'aid' provision of 19 million EUA and for Suriname (population 389,000 and GNP *per capita* US$2,110) some 18 million EUA.[52] France was equally to favour its Caribbean overseas territories, which could also draw, because of their status as *départements*, on the EEC Regional Development Fund. The LDCs of the Commonwealth Caribbean, in particular those in the eastern Caribbean, saw this as grossly unfair. They put their case forcefully to Maurice Foley, Minister of State at the Foreign and Commonwealth Office (now Deputy Director General of the Development Division of the EEC) at a meeting in Antigua. Britain, however, was powerless to change the formula and Foley was subsequently to offer only the smallest crumb of comfort in a letter pointing out 'the total volume of EEC aid for [dependencies] as a whole is far higher per head of population than that available to the ACP states'.[53] That in practice this was to amount to very little is easily seen by comparing the provision made for Grenada and that for the other eastern Caribbean islands.

The 'aid' made available to the LDCs totalled 19,660,000 ECU. Again a degree of equity was ensured, with allocations among individual territories as follows: Antigua, 2,080,000 ECU; St Kitts-Nevis, 1,590,000; Dominica, 2,500,000; St Lucia, 3,215,000; St Vincent, 3,060,000; Grenada, 2,000,000; Belize, 3,815,000; other Caribbean, 1,400,000.[54] However, with respect to commitment by sector a marked difference in pattern from the MDCs is seen, with the LDCs focusing overwhelmingly on economic infrastructure (except for St Lucia).[55] This stems in part, as noted earlier, from the British bilateral funding of the principal 'development of production' programme in the banana development scheme. It does not explain it all, however, and uniformity is likely to have another cause than merely identical need. On this matter the Commission itself is adamant, arguing, indeed, that it was obliged to accept 'last resort' projects, some of which were rather suspect.[56] If this is so, then direction must have come from the British government via the British Development Division in the Caribbean, which retained overall responsibility in this area. Finally, as with the MDCs, the commitment rate is high (over 80%) with the sole exception of Antigua. No satisfactory explanation is available to account for this, though difficulties of supply for projects have been encountered.[57]

The third head under which EDF funds are distributed is the Caribbean regional programme. The Lomé Convention identified regional co-operation as an area deserving particular support and 10% of EEC funds were set aside for financing regional projects. This emphasis was in accord with the desires of Commonwealth Caribbean states, who through the Caribbean Development Bank and the CARICOM Secretariat were in a strong position to take advantage of them.[58] Thus while the amount finally allocated of some 26 million EUA was less than half that sought, a number of projects identified, especially in infrastructure, were to be agreed. These are set out in Table 8.3.

Other interventions. 'Other interventions' in total amount to just under 50% of the 'aid' allocated under the Fourth EDF. They are thus by no means an inconsiderable aspect of EEC–Commonwealth Caribbean relations.

The largest amount committed is some 25 million ECU via the EIB. Here the MDCs have benefited considerably, as the following shows: Trinidad and Tobago, 10,000,000 ECU; Barbados, 7,500,000; Guyana, 3,200,000; St Lucia, 180,000; Jamaica, 70,000; regional, 4,000,000.[59] The sums for Trinidad and Tobago, for the region, and one third of that committed to Barbados, have gone towards the promotion of small and medium-sized business. This is in line with the EIB's own preferred emphasis on support for this sector within its overall obligations. It also corresponds, in the case of Trinidad and Tobago, with a conscious emphasis on EIB financing as against the EDF, the latter, with its cumbersome bureaucratic procedures, being reserved largely for the provision of technical assistance.[60]

The next largest amount committed has been 'food aid'. The beneficiaries, in descending order, are: Jamaica, 12,402,000 ECU; Guyana, 4,507,000; Grenada, 1,722,000; Dominica, 830,000; Antigua, 689,000; St Lucia, 206,000.[61] The aid has consisted principally of cereals, milk powder and butter oil, the latter two constituting, of course, the classic agricultural 'surpluses' of the EEC.

'Exceptional aid' has amounted to nearly 8·5 million ECU. This has been given following hurricanes David and Frederick in 1979, Allen in

1980, the eruption of Mount Soufrière in St Vincent in 1979, and torrential rainfall in the Windward Islands and Jamaica. It has essentially, then, been 'emergency' or 'disaster' aid and has, in consequence, mainly taken the form of tents, blankets, medicine, building materials and agricultural inputs. The chief beneficiaries (but correspondingly worst affected) have been Dominica, 4,400,000 ECU; Jamaica, 1,175,000; St Lucia, 1,000,000; St Vincent, 1,000,000; Grenada, 450,000.[62] Given the amount of damage caused, this aid has been particularly welcomed, especially by the LDCs.

Table 8.3 *Regional co-operation: Fourth EDF*

Projects	Commitments ('000 ECU)
West Indies Shipping Corporation	6,373
Common Services (ECCM)	1,200
CDB (technical assistance)	440
CDB (studies)	1,060
Corn project (Belize)	1,925
Regional fisheries	14
University of West Indies	3,000
Caribbean Agricultural Research and Development Institute	1,500
Caribbean Tourism Research Centre	538
Caribbean Food Corporation	300
CARICOM	200
Leeward Islands Air Transport[a]	5,420
Caribbean Aviation Training Institute	2,300
Training	670
Ferry, Guyana–Suriname[b]	1,000
Total	25,940

Notes.
(a) 5,245 ECU to be committed.
(b) 1,000 ECU added from Suriname.
Source. Commission of the European Community, *La CEE et la région des Caraïbes Lomé I et Lomé II*, Brussels, 15 February 1982.

One final point must be made about 'aid' – the cost and complexity of administration. Under Lomé I the cost of EEC delegations was borne direct by the host ACP state as deductions from the EDF grant allocations. This was considered offensive by many in the ACP, particularly when the costs of the delegations were high. This would certainly appear to be the case in the Commonwealth Caribbean, where the costs of delegations for 1976–79 reached 5,261,000 EUA.[64] Against this, though, must be set the fact that multiple accreditation is a feature of all Commonwealth Caribbean EEC delegations and so costs are easily inflated. This is further confirmed from personal experience of visits to delegations, where most appear over- rather than underburdened with work. This, in itself, apart from procedure, must account for some delay in aid provision, as evident in the Commonwealth Caribbean as elsewhere. One further point to be noted in this regard is a general lack of co-ordination between agencies. For example, the delegation of the EEC in Barbados has for much of its time been ignorant of the parallel work of the British Development Division in the Caribbean, itself located in Barbados, and vice versa. At the other end, the direct lines of communication between the Commission in Brussels and its delegations in the Caribbean considerably reduce the role of some Commonwealth Caribbean missions in Brussels, keeping them in ignorance even of the general progress of the aid programme.[65]

Industrial co-operation. The Lomé Convention has a separate chapter on industrial co-operation. This was added largely on the initiative of the ACP states and is loosely based on the memorandum they submitted on this subject at the Kingston Ministerial conference. It therefore represents an example of the influence of the ACP states in shaping the Convention and is a significant departure from the casual attitude previously shown by the EEC toward promoting industrialisation.[66]

As set out in the Convention, industrial co-operation was to promote the development and diversification of industry in the ACP states; promote relations with the Community; increase the links between industry and other sectors of the economy, including agriculture; facilitate the transfer of technology to the ACP states; promote the marketing of industrial products; and encourage small and medium-sized firms and Community firms to participate in the industrial development of the ACP states. These objectives were to be realised by funds made available through the EDF and EIB and through specific institutions designed for the task, viz. a Centre for Industrial Development (CID) overseen by an Industrial Co-operation Committee.[67]

This new emphasis on industry was, of course, particularly welcome to the Commonwealth Caribbean. Development plans had, for many years, lauded the virtues of industrialisation, and all governments were committed to raising considerably the share of manufacturing in GDP. This said, several qualifications must be noted at the outset in terms of constraints on the industrial programme in the Commonwealth Caribbean deriving from the general strategy Lomé I proposed. Essentially, this emphasised the 'voluntary' nature of most industrialisation, reflecting the belief of EEC governments that market forces would be the most significant factors influencing industrial co-operation. In the Commonwealth Caribbean such market forces are directed overwhelmingly to and from North America. Private investment increasingly comes from the USA and there is little reason to suppose that the EEC could or would match it in aggregate. Trade, as we have already seen, shows a similar pattern. In short, development of production in close co-operation with EEC firms was unlikely to materialise.

Industrialisation for the Caribbean within the context of Lomé was concerned mainly with access to the EEC market. Here the restrictive definition of rules of origin adopted under Lomé I was a particular limitation on development, given the high import content of manufacturing within the region as a whole. To this, naturally, was added the question of transport costs. Again constraints, rather than opportunities, appear to characterise possibilities, weakening the impact of the programme as a whole. Not much, then, was expected, and not much transpired. What little was achieved may be appreciated under the broad headings of 'funds for industrialisation' and 'institutional interventions'.

Funds for industrialisation, properly so called, totalled only some 19 million ECU. Of this, the MDCs received nearly all, with the largest single amount allocated to Guyana (9·2 million EUA) followed by Trinidad and Tobago (6 million EUA) and Barbados (2·9 million EUA).[68] The LDCs, on the basis of their national indicative programmes, hardly figured at all, though it should be noted that the 4 million ECU loan to the Caribbean Development Bank has been given on the understanding that promotion of industrial development among these territories will be a priority. As with the 'aid' programme, then, and as an integral part of it, the Commonwealth Caribbean as a whole has been disadvantaged. Of the 222·2 million ECU committed for industrial projects among the ACP under the Fourth EDF the Commonwealth Caribbean received only 5·7 million ECU (2·5%) – and that wholly in regard of Guyana.[69]

With so little finance directly committed, the institutional interventions take on an added significance. Under Lomé I, however, neither the CID nor the Committee for Industrial Co-operation has had a particularly distinguished record. The former, for instance, was not established until January 1977, while the latter did not give the CID guidelines for the conduct of its activities until October 1978. Additionally, both staff and budget were severely limited[70] and in no sense adequate to carry out the tasks of the CID as set out in the Convention itself. However, what little was on offer the MDCs of the Commonwealth Caribbean have proportionately sought to utilise. The record of CID interventions to the end of December 1980 illustrates this. Of a total of 471 national/regional interventions for all ACP countries the Commonwealth Caribbean account for forty-six (of which thirteen were from Trinidad and Tobago and twelve from Barbados) and of a total of ninety-nine studies co-financed no fewer than thirteen.[71] The focus of such interventions, as to be expected, has been the promotion of light industry and the provision of technical assistance.[72] Here, as elsewhere, the final results in terms of projects successfully completed, remain disappointing.

Institutions. Under the earlier association arrangements of the AASM with the EEC, institutions had been established but had not played a significant part. With Lomé I, however, this was to change as a range of joint ACP/EEC institutions were established on the principle of parity and dialogue at a number of levels, this to be reflected at the same time within a separately constituted ACP group. In the negotiation for the Convention the Commonwealth Caribbean showed far more interest in the possibilities of the latter than the former, thereby demonstrating again their basic worries and suspicion of arrangements that might be construed, or in practice prove to be, neo-colonial, while enthusiastically endorsing any manifestation of South—South co-operation. In the event their aspirations were not to be realised, as the joint institutions proved more effective and the ACP machinery less useful than once thought. The separation of institutions remains a feature, though, and may be so considered below.

Joint institutions. The principal institutions of the Convention are the Council of Ministers, assisted by the Committee of Ambassadors and the Consultative Assembly. Continuity is assured by delegated powers, sub-committees and executive agencies of the above.

The Council of Ministers is the supreme organ of the Convention and composed of the members of the Council of the EEC and members of the Commission on the one hand, and a member of the government of each ACP state on the other. In the five-year period 1976–80 the Council of Ministers has met five times, twice in Brussels and once each in an African, Caribbean (the Bahamas, March 1979) and a Pacific state. It has also met on several occasions as a negotiating conference in respect of the renewal of the Lomé Convention. Decisions taken by the Council in 1978 have had the effect of strengthening the role of the ACP–EEC Committee of Ambassadors, especially in matters with a large administrive content. This has left the Council to concentrate on issues which are basically political and of which sugar has been the most directly important to the Caribbean. The rotation principle of the office of president has also allowed the Commonwealth Caribbean an input – G. King (Guyana), P. J. Patterson (Jamaica) and B. St John (Barbados), all having held office.

The ACP-EEC Committee of Ambassadors consists, on the one hand, of one representative from each EEC member state and of a Commission representative, and on the other hand, of a representative from every ACP state.[73] It is assisted by three committees and five sub-committees, the former including a Permanent Joint Group for Bananas and the latter a Sub-committee for Sugar. From 1976 to 1980 the full Committee of Ambassadors met only nine times outside the framework of the renewal of the Convention. By contrast the sub-committees met far more frequently, for example fifty-seven meetings took place between April 1976 and July 1978.[74] This attests to the importance of the latter in maintaining the momentum of the Convention, and the work of several sub-committees stands out in particular, notably sugar and commercial co-operation.[75] Here it is more than of passing note that the interests of the Commonwealth Caribbean are vitally bound up with the effective functioning of these two committees and that the chairmanship of them has fallen to relatively dynamic Commonwealth Caribbean ambassadors.

The Consultative Assembly consists on the one hand of members of the European Parliament, and on the other of representatives appointed by the ACP states, both parties appointing the same number. It meets annually and its principal function has been to express opinions in the form of resolutions on matters covered by the Convention. To aid it in this task a Joint Committee has been established, again on the basis of parity, and from which a number of important reports have originated.[76] While it has no real powers (perhaps because of this) the

Consultative Assembly, and especially the Joint Committee, have unexpectedly turned out to be the most dynamic institutions created by the Lomé Convention. This has been particularly marked since 1979, and all Commonwealth Caribbean spokesmen interviewed by the author in Brussels have endorsed the activities of the Assembly and the Committee as the most truly expressive of the 'spirit' of the Convention. Certainly the resolutions and reports of the Assembly have been supportive of ACP–EEC co-operation, often with an emphasis on the ACP over the EEC. The Assembly has also been active in 'wider' areas, notably in pressing for action against South Africa. Finally, while the Consultative Assembly has always met in Luxembourg, the Joint Committee has made a practice of meeting in various ACP countries, on one occasion in the Commonwealth Caribbean (Grenada, May 1978).

ACP institutions. The Commonwealth Caribbean in general, and Ramphal in particular, did much to encourage and sustain the idea of co-operation among the ACP states as a corollary to the negotiations of Lomé I. Once agreement with the EEC was reached it was therefore only fitting that Guyana should host the conference that institutionalised this arrangement in the Georgetown Agreement, adopted on 6 June 1975. This set out the aims and objectives of the ACP group and specified the organs by which they were to be realised. Simply stated, these were to 'deepen' existing solidarity through the establishment of parallel institutions to those provided within the Convention itself. Thus an ACP Council of Ministers, an ACP Committee of Ambassadors and an ACP Secretariat were established. In turn it was expected that these, in their attempt to ensure effective implementation of the Convention, would not only promote greater South–South co-operation among themselves but also, by force of example, further the establishment of the NIEO.[77]

The ACP Council of Ministers consists of one member of the government of every ACP state or a representative appointed by him. It is more active than its joint counterpart and to date has met twice in the Commonwealth Caribbean – once in the Bahamas (March 1979) and once in special session in Jamaica in July 1980, when the discussion, taking its cue from the Manley government, strongly emphasised the consolidation and extension of ACP co-operation.[78] The presidency has been held four times by the Caribbean – G. King (Guyana), P. J. Patterson (Jamaica), B. St John (Barbados) and H. Shearer (Jamaica). The fact that Trinidad and Tobago has not held the presidency is due to its relative indifference at this level to the Lomé Conventions.

The ACP Committee of Ambassadors consists of an ambassador and a representative from each ACP state. As with the ACP Council, it is far more active than its joint counterpart. It meets at least once a month in full session and sub-committees usually more frequently. This level of activity makes strenuous demands on all involved, and the principal function of the Commonwealth Caribbean embassies in Brussels appears to revolve around servicing it and its various committees.[79] The record of such committees, as to be expected, is mixed, but once again the sugar sub-committee stands out as particularly active.[80] To date the Commonwealth Caribbean has supplied three chairmen to the full committee – Dr J. O'Neil Lewis (Trinidad and Tobago, 1976), D. Rainford (Jamaica, 1979), and O. Jackman (Barbados, 1982).

The decision to set up an ACP Secretariat was an integral part of the Georgetown Agreement. The Secretariat was to be located in Brussels and its chief functions were to monitor the implementation of the Lomé Conventions and to assist the ACP Council and ACP Committee of Ambassadors. The Secretariat was established in 1976 and currently has a staff of fifty-one. The two most important posts are those of Secretary General and Deputy Secretary General. The former has been held by two Africans and the latter by a Trinidadian, Edwin Carrington. At the next level three persons from the Caribbean have filled senior posts, with two other persons of Caribbean origin further employed at lower levels. This somewhat skewed staffing pattern clearly demonstrates the relative educational advantage of the Commonwealth Caribbean over the African and Pacific states and further underlines the technocratic input emanating from the region which has been so characteristic of its contribution to Third World development.

The Commission. Finally, mention must be made of the Commission. It remains the most important organ in ACP–EEC relations. Within the Commission a special division, Directorate General VIII (DG8), is responsible for development, aided by a number of delegations overseas. Internally within DG8 matters wholly relating to the Caribbean are handled by a specific desk (the same officer is also concerned with Indian Ocean and Pacific Ocean affairs), though, by definition, there is considerable overlap both with DG8 and with other general directorates. On financial and technical co-operation, in particular, lines of communication go direct from DG8 to the various delegations in the Caribbean, and elsewhere, for onward transmission

Britain and the EEC in the Commonwealth Caribbean

to respective governments. The overseas delegations therefore play a critical role in this field. Within the Commonwealth Caribbean there are four. They are in Barbados (also responsible for Antigua, Dominica, St Lucia, St Vincent, St Kitts-Nevis, Montserrat, Anguilla, the British Virgin Islands, and the Caribbean Development Bank); Guyana (also responsible for the CARICOM Secretariat); Jamaica (also responsible for the Bahamas, Belize, the Cayman Islands and the Turks and Caicos Islands); and Trinidad and Tobago (also responsible for Grenada, Martinique, Guadeloupe and dependencies, Guiana, St Pierre and Miquelon, and the South Atlantic territories). The responsibilities of the various delegations are thus extensive. It is by no means certain that in every case they have the resources to carry them out effectively.[81]

The negotiations of Lomé II: Caribbean concerns

The negotiations for Lomé II formally opened on 24 July 1978. They reflected, in the opening statements of both parties, what was already well known – that for the EEC the emphasis was on *renewal* of the existing Convention whereas for the ACP it was on *renegotiation*.[82] This substantive difference was to make the negotiations both protracted and difficult, and a successful conclusion was not reached until 31 October 1979. The details of the negotiations are once again of no immediate concern.[83] Open to consideration, however, is whether the Commonwealth Caribbean demonstrated a specific interest in the negotiations as an identifiable group; and whether, under Lomé II, benefits were likely to be greater than under Lomé I.

In respect of the first question, several observers have argued that the Commonwealth Caribbean had nothing really significant at stake in the negotiations. The sugar protocol was not up for renegotiation, and the Commonwealth Caribbean's interest in trade and particularly 'aid' was less than that of either the African or the Pacific states.[84] There is some truth in this, as the record of Lomé I clearly shows. It was also, though, the product of a conscious decision which had two elements. One was that now three Commonwealth Caribbean MDCs were represented in Brussels by permanent delegations, with general interests being serviced by an ACP secretariat, there was no need for specific intervention. Correspondingly, the CARICOM Council of Ministers took the decision that the role of the CARICOM Secretariat in negotiations would be 'minor'.[85] The second was that by a fortuitous combination of circumstances the Commonwealth Caribbean would occupy key chairmanships both at the beginning and during the crucial middle part of negotiations, thereby allowing any specific intervention to be pursued under the guise of the general interest. As luck would have it, a further Commonwealth Caribbean figure, in the shape of B. St John, was also to conclude negotiations, providing yet another and this time unanticipated bonus.

If the Commonwealth Caribbean was not inclined to intervene in negotiations for 'regional' ends[86] it was nevertheless well placed to set both the tone and the agenda. This it proceeded to do at the Ministerial level, with P. J. Patterson, first as ACP president and then as ACP chairman of the trade and customs co-operation sub-committee, complemented at the ambassadorial level by D. Rainford (Jamaica). Under this arrangement Patterson vigorously promoted the extension and consolidation of duty-free access to the EEC, paying particular attention to the 'safeguard clause'.[87] He also argued strongly that the relationship of the ACP to the EEC was one of interdependence, not dependence,[88] a theme taken up by Rainford with his assertion that co-operation was not one-sided, 'the ACP states receiving without giving. Europe needs the ACP countries as much as the latter needs the former.'[89] It was this that was to be the major difference of perspective, not only with the EEC but also with a number of the francophone African states who adopted a supplicatory attitude to negotiations as a whole. And here, if anywhere, lay the source of the Commonwealth Caribbean's reputation for militancy, one which was further confirmed, if somewhat less stridently, when B. St John assumed a leading role in the final stages of the negotiations and attempted to resolve by forceful argument alone the outstanding issue of investment agreements.

The Commonwealth Caribbean's presence in the negotiations was thus tangible. It was recognised by the EEC as such, notably in the dispatch of a high-level Commission delegation led by the head of DG8, Claude Cheysson, to key African and Caribbean capitals following the breakdown of negotiations in May 1979. This initiative did succeed in bringing the two sides together at the end of June, and thereafter negotiations continued until agreement was reached. However, the 'spirit of partnership' lauded in 1975 was not 'rekindled'[90] and the final outcome was barely satisfactory to the ACP as a whole. B. St John emphasised this in his speech at the signing ceremony, when he spoke of 'long and arduous' negotiations which left 'a sense of frustration' among ACP states that 'legitimate demands and requirements for economic survival and advancement' remained 'in urgent need of sympathetic responses'.[91] In short, renewal had triumphed over renegotiation, and continuity was more evident than change.

Given this outcome, what did the Commonwealth Caribbean stand to gain? Taking the principal headings of the earlier assessment of Lomé I the answer must be: very little.[92]

Commercial co-operation. The ACP demands for free access for all agricultural products, for greater liberalism in the application of the rules of origin for semi-processed goods, and for a much circumscribed application of the safeguard provisions by the EEC, were not met in full. Some points were conceded, but only in very small part, and the improvements were at best only marginal. The same applies to the operation of the rum and banana protocols, with the former benefiting from only a limited increase in duty-free quotas on continental European markets and the latter from minor institutional improvements designed to promote production and marketing. Trade promotion was to be boosted and Stabex extended and improved, with dependence and trigger thresholds lowered and reimbursement terms eased. Both these provisions were potentially of limited benefit to the Commonwealth Caribbean, whilst major demands such as the inclusion of tourism within the ambit of Stabex were resisted. Only in one area was there innovation – the provision of an accident insurance scheme for ACP mineral producers known as Minex. This has basically the same aim as Stabex, and the inclusion of bauxite and alumina as minerals to which the scheme will apply is clearly of significant benefit to Caribbean states. The indications are that Guyana has sought, or will shortly seek, assistance under this provision.[93]

Financial and technical co-operation. The resources available under this head have allegedly been increased by 62% to 5,227 million EUA. As Hewitt and Stevens have pointed out, however:

> This figure is totally misleading. Not only does it ignore inflation and the swelling of the ACP ranks but it also includes funds that do not qualify as aid (official development assistance) or even as hard loans as well as some figures that are purely cosmetic. When account is taken of all these factors, the real oda per head of the ACP population has been cut by around one-fifth.[94]

For the Commonwealth Caribbean, figures given by the indicative programmes suggest cuts far greater than this.[95] The increase in the MDCs' indicative programmes in money terms is only of the order of

5% minimum to 29% maximum, and for the LDCs from 34% minimum to 44% maximum. This represents a real cut which increases in regional aid of the order of 200% can do little to offset. Furthermore the real benefits accruing to the Commonwealth Caribbean 'island state' and/or 'least developed state' category under Lomé II[96] have in this case been counterbalanced by the Commission's stated policy of taking into account the budget of the sugar protocol and 'aid' given under other EEC policies in assessing the full extent of EEC financial and technical assistance. Either way the Commonwealth Caribbean is the loser, and its previous strong advocacy for a say in the administration of aid, in part now conceded, is but small recompense.

Industrial co-operation. In voicing his disappointment with the outcome of negotiations B. St John remarked, 'Probably the greatest fear of the ACP is the lurking danger of enduring another five years with a chapter on industrial co-operation which could find itself devoid of operational content due to lack of adequate financial resources.'[97] The ACP has sought to counter this limitation with its proposal for an Industrial Development Fund, but it found no favour, and all the EEC was prepared to do was to strengthen existing arrangements. The CID and EIB thus remained the focus of industrialisation efforts buoyed by greater funds, in the case of the Commonwealth Caribbean MDCs an increase in money terms of some 88%.[98] Efforts to promote the development of mining and energy projects are also areas of potential benefit to the MDCs. The Commonwealth Caribbean LDCs, as before, are likely to benefit only marginally, if at all. Their greatest gains (if any) are likely to be made through the emphasis on agricultural co-operation in Lomé II and specifically through the establishment of a Technical Centre for Agricultural and Rural Co-operation.

Institutions. No major changes were introduced here, though greater flexibility within and between the various organs was to be encouraged.

New areas of co-operation. Joint declarations on migrant workers, on fishing, and on shipping are all of potential significance to the Commonwealth Caribbean. They remain, of course, to be proved in practice.

It can therefore be seen why, in EEC publications describing Lomé II, 'extension' and 'consolidation' are the most frequently encountered terms and 'resignation' is the most frequently encountered ACP response. The gap between aspiration and achievement remains large, and already from the Commonwealth Caribbean has come the warning of 'discernible limits to the progress we can make in this area'.[99] In short, the development thrust of the Lomé Convention leaves much to be desired. What, then, is the nature of the relationship that has come to exist between the Commonwealth Caribbean and the EEC?

From neo-colonialism to neo-colonialism?

Reflecting on his considerable experience as Trinidad and Tobago's ambassador to the EEC, Dr O'Neil Lewis has written:

> The provisions of the Lomé Convention provide some ... valuable assistance, because no other assistance of that nature is available. But it is no more than a step in the right direction. And, on sober reflection, it will be seen as a step that may do only a little good but certainly does no harm either to the ACP or to those countries, those developing countries, for which there are no arrangements comparable to the provisions of Lomé. What is needed is not merely the extension of Lomé benefits to as many other developing countries as possible, but the adoption by other industrialized countries – the USA, the USSR, Japan, Canada, for example – of Lomé-type agreements for the benefit of the developing world.[100]

For the Commonwealth Caribbean Lomé is very much a second best. It is the product of an historical relationship with Britain that has been multilateralised by British entry to the EEC but not changed in essence by that fact. Commercial co-operation predominates, and within that sugar remains critical. All other relationships with the EEC are but alien grafts upon this stock, yet to take hold let alone to flower!

In the final analysis, then, at least for the Commonwealth Caribbean, Lomé appears more to preserve the past within a new set of arrangements than to prefigure the future. If it is a change it is from one set of arrangements which may be identified as 'neo-colonial', i.e. post-independence relations with Britain, to another set of arrangements which in themselves can attract that label, i.e. indirect and subtle domination by political, economic and technical means. If this is a contentious point, then all that can be said is that it is so perceived by those who have to deal with the matter daily. In answer to the question 'What is the essence of the relationship of the EEC to the Caribbean?' all EEC officials interviewed replied, *Politics*, all Commonwealth Caribbean officials *Economics and politics*. In the complex of international relations in which the Commonwealth Caribbean is necessarily drawn, and in particular in its relations with the USA, this is a fact to be kept prominently in mind.

Notes

The author gratefully acknowledges the assistance of a grant from the Nuffield Foundation for fieldwork in the Commonwealth Caribbean and in Brussels, 1981–82.

1. See especially *The Courier* (Special Issue), No. 31, March 1975.
2. Issued in May 1970, this was a revised edition of a pamphlet first published in December 1967.
3. Associated Overseas Territories (AOT) status was established for dependent territories by Part IV of the Treaty of Rome and for independent territories by the first Yaoundé Convention. It provided for the progressive establishment of a free-trade area between the EEC and the associated countries and territories, by the reciprocal reduction of tariffs and quantitative restriction; and the establishment of a European Development Fund for the purpose of granting EEC financial aid to the associated countries and territories.
4. Mr Geoffrey Rippon in the House of Commons, 17 May 1971 (Hansard, cols. 385–6) cited in Central Office of Information, *Britain and the Developing Countries: The Caribbean*, London, HMSO, 1973, p. 21.
5. 'Statement from the Lancaster House Conference' cited *ibid*. Significantly, when Rippon circulated the Lancaster House declaration to the EEC Ministers of Foreign Affairs during the course of a meeting in Luxembourg on 7 June they took note only of the declaration, and the text did not even appear in the minutes of the meeting.
6. For details see *Treaty Concerning the Accession of the Kingdom of Denmark, Ireland, the Kingdom of Norway and the United Kingdom of Great Britain and Northern Ireland to the European Economic Community and the European Atomic Energy Community*, Cmnd 4862-1, London, HMSO, January 1972. This was virtually identical to the offer made by the EEC Commission in the 'Declaration of Intent' in 1963.
7. Source: interview.
8. C. Cosgrove Twitchett, *A Framework for Development: The EEC and the ACP*, London, 1981, p. 9.
9. Source: interview.
10. Figures from *The Courier*, No. 31, p. 12.
11. Hon. Shridath S. Ramphal, Minister of Foreign Affairs, Guyana, *Just, Enlightened and Effective Arrangements: New Approaches to Relations with the European Economic Community* (Statement delivered at the Opening Conference on Behalf of the Delegation of Caribbean Countries, 26 July 1973), Georgetown, July 1973.
12. This was 'unclear' to the EEC until the very last moment, the Caribbean

on the day sitting together behind a hastily written notice announcing 'the Caribbean' instead of seating themselves individually by state as expected.
13 *The Courier*, No. 31, p. 23.
14 Two useful accounts are C. Webb, 'Mr Cube *v*. Monsieur Beet: the politics of sugar in the European Communities' in H. Wallace *et al*. (eds), *Policy Making in the European Communities*, Chichester, 1978; and Europe News Agency *Sugar: Europe's New Policy*, Brussels, 1980.
15 Source: interview, reporting a meeting between P. Southwell, representing the Associated States, and senior officials of the Foreign and Commonwealth Office in London, 15 September 1972.
16 Source: interview.
17 See N. A. Adams, 'The Lomé Convention and Caribbean trade' in L. F. Manigat (ed.), *The Caribbean Yearbook of International Relations 1976*, Sijthoff, Leiden; IIR-UWI, Trinidad and Tobago, 1977.
18 For details see Eurostat, *Analysis of Trade between the European Community and the ACP States*, Luxembourg, 1979.
19 Figures calculated from *La CEE et la région des Caraibes Lomé I et Lomé II*, CEE/DGD, Direction B, Caraibes, Brussels, 15 February 1982, internal memorandum.
20 *Ibid.*; and calculated from Eurostat, *ACP: Yearbook of Foreign Trade Statistics 1972–1978*, Luxembourg, 1981.
21 *Ibid.*
22 Sources for Grenada as in n. 20; for St Lucia inferred from Government of St Lucia, *Annual Statistical Digest, 1978/1979*, Castries, St Lucia; for St Kitts-Nevis inferred from Government of St Kitts-Nevis, *Annual Digest of Statistics 1979*, Basseterre, St Kitts-Nevis.
23 *Ibid.*
24 Adams, 'The Lomé Convention and Caribbean trade', pp. 261–62.
25 Cited in European News Agency, *The Lomé Convention: Renegotiation and Renewal* (study written under the direction of J. A. Fralon), Brussels, 1978, p. 121.
26 Source: interviews.
27 Figures derived from Europe Information, *Sugar, the European Community and the Lomé Convention*, Brussels, 1979.
28 Source: interviews.
29 For a useful statement on this see European News Agency, *Sugar: Europe's New Policy*, chapter 4, especially pp. 136–40.
30 The European Currency Unit (ECU) is a composite monetary unit consisting of a fixed 'basket' of currencies of the member states of the EEC and is equivalent to the European Unit of Account (EUA). Its value is calculated daily by reference to the market exchange rates of the component currencies. On 1 October 1980 it was assigned a value 1 EUA/ECU = US$1·405.
31 For the precise wording of the protocol see ACP-EEC Convention of Lomé (complete text) in *The Courier*, No. 31.
32 A review of the various national regimes is to be found in Europe Information, *Bananas: Essential Elements of the World and*

Community Markets, Brussels, April 1978.
33 Adams, 'The Lomé Convention and Caribbean trade', p. 268.
34 Figures calculated from Ministry of Overseas Development, United Kingdom, *Report of the First ODM Annual Review of the Windward Islands Five Year Banana Development Programme*, mimeo, March 1979.
35 *Ibid*.
36 Chris International Foods Ltd, a West London firm, claim that the allocation of dollar banana licences by the UK Ministry of Agriculture constitutes a monopoly and is contrary to the Treaty of Rome. The case is currently before the courts.
37 ODM, *Report of the First ODM Annual Review . . . Windward Islands*.
38 According to one account the British government meant this figure to be 14% and not 40%. The result of a misunderstanding/mistranslation of *quatorze* and *quarante*? See C. Cosgrove-Twitchett, 'From association to partnership' in K. J. Twitchett (ed.), *Europe and the World: The External Relations of the Common Market*, London, 1976, n. 22, p. 139.
39 Figures from *La CEE et la région des Caraibes Lomé I and Lomé II*.
40 Source: interviews and report of the ACP-EEC Council of Ministers (1 April 1976–29 February 1980), *ACP-EEC Co-operation: Analysis–Application*, Brussels, 25 July 1980, pp. 49–57.
41 A useful account of Stabex is to be found in the *ACP States Yearbook, 1980–1981*, London, 1980, pp. 609–17.
42 OCT is the formal designation given within the EEC to the dependent countries and territories associated under Part IV of the Treaty of Rome, as against the independent countries associated under the Lomé Conventions (ACP).
43 Figures from *The Courier*, No. 69, September–October 1981, pp. 50–1.
44 Figures from *La CEE et la région des Caraibes Lomé I et Lomé II*.
45 Source: interviews.
46 Figures from *The Courier*, No. 31.
47 Figures calculated from *La CEE et la région des Caraibes Lomé I et Lomé II*.
48 Source: interview.
49 Figures from *La CEE et la région des Caraibes Lomé I et Lomé II*.
50 Figures from *ACP-EEC Co-operation: Analysis–Application*.
51 Cosgrove Twitchett, *A Framework for Development: The EEC and the ACP*, p. 64.
52 Figures for aid from *La CEE et la région des Caraibes Lomé I et Lomé II*.
53 Cited in A. E. Thorndike, 'The Concept of Associated Statehood with special reference to the Eastern Caribbean', unpublished PhD thesis, University of London, 1979, p. 269.
54 Figures from *La CEE et la région des Caraibes Lomé I et Lomé II*.
55 Calculated from Delegation of the Commission of the European Communities in Barbados, *Lomé Convention: 4th EDF. National Indicative Programmes – Project Profiles*, mimeo, June 1981, except for

Britain and the EEC in the Commonwealth Caribbean 235

Grenada, from *ACP-EEC Co-operation: Analysis–Application*.
56 Source: interview.
57 Source: interview.
58 The CDB was to note that the definition of regional projects in Lomé 'conforms closely to the Bank's conception of the issue'. Caribbean Development Bank, *Proposals regarding Regional Projects prepared for EEC Programming Mission, 24 September 1975*, mimeo, p. 8.
59 Figures from *La CEE et la région des Caraibes Lomé I et Lomé II*.
60 Interview with Frank Barsotti, Permanent Secretary of the Ministry of Finance, *The Courier*, No. 52, November–December 1978, pp. 34–7.
61 Figures from *La CEE et la région des Caraibes Lomé I et Lomé II*.
62 *Ibid.*
63 Other 'aid' given has been in respect of non-governmental organisations and to Guyana. The former has seen disbursements of 'aid' to Jamaica, Dominica, Grenada, Guyana and Trinidad and Tobago in that order. The latter has been a 'European' contribution to an 'aid' package constituted by the World Bank, OPEC and other industrialised countries to aid Third World countries most affected by the oil crisis. In the case of Guyana the aid has taken the form of fertiliser for Guysuco and spare parts for the bauxite industry.
64 Figures from *ACP-EEC Co-operation: Analysis–Application*. The individual totals are as follows: Barbados, 1,232,000 EUA; Guyana, 914,000; Jamaica, 1,694,000; Bahamas, 49,000; Trinidad and Tobago, 1,345,000; Grenada, 27,000.
65 Source: interviews.
66 The Convention of Application annexed to the Treaty of Rome allocated only 1% of finance to industrial undertaking. This was increased to 12·7% in the first Yaoundé Convention but fell back to only 10·8% in the second. Figures from *ACP States Yearbook 1980–1981*, p. 629.
67 For details see ACP-EEC Convention of Lomé, Title III, in *The Courier*, No. 31.
68 Figures from *ACP-EEC Co-operation: Analysis–Application* (as at 31 December 1979).
69 See *The Courier*, No. 60, March–April 1980.
70 A professional staff of ten (of whom only three were full-time) were allocated a total budget, 1977–79, of only 6,389,466 EUA. See *ACP-EEC Co-operation: Analysis–Application*.
71 Figures from Centre for Industrial Development, *Annual Report 1980*, Doc. CIC-CID/5/81, Annex.
72 For details see Centre for Industrial Development, *ACP-EEC Industrial Co-operation: Interventions* (file contents at 28 February 1980).
73 A representative of the EIB is present at meetings of the committee when its agenda includes questions falling within the scope of the EIB's activities.
74 Figures from *ACP-EEC Co-operation: Analysis–Application*.
75 Source: interviews.
76 See in particular Katharina Focke, *From Lomé I towards Lomé II*

(Texts of the report and resolution adopted on 26 September 1980 by the ACP-EEC Consultative Assembly), Luxembourg. Ambassador Insanally (Guyana) prepared the report for 1981.

77 For details see *The Courier*, No. 33, September–October 1975.
78 For details see *The Courier*, No. 63, September–October 1980.
79 As a result of the 'geographic' share-out of committees the Caribbean has to provide eight vice-chairmen, and the work of the committees is such that this can often be an onerous post.
80 Source: interviews.
81 Source: interviews.
82 Compare the contrasting statements of Hans-Dieter Genscher, then President of the EEC Council of Ministers, with that of P. J. Patterson, then President of the ACP Council of Ministers. See *Speech delivered by the President of the Council of the European Communities on the occasion of the opening ceremony of the negotiations for the new ACP-EEC Convention*, mimeo, and *Statement by the President of the Council of ACP Ministers on the occasion of the opening of the negotiations for the successor arrangement to the Lomé Convention*, mimeo.
83 Useful overviews of negotiation and outcome are Cosgrove Twitchett, *A Framework for Development: The EEC and the ACP*; John Ravenhill, 'Asymmetrical interdependence: renegotiating the Lomé Convention' in F. Long (ed.), *The Political Economy of EEC Relations with African, Caribbean and Pacific States*, Oxford, 1980; and Adrian Hewitt and Christopher Stevens, 'The second Lomé Convention' in C. Stevens (ed.), *EEC and the Third World: A Survey, 1*, London, 1981.
84 Cosgrove Twitchett, *A Framework for Development*, pp. 102–3; Ravenhill in Long, *The Political Economy of EEC Relations*, pp. 38–9.
85 Source: interview. Under Annex V to Lomé I the Caribbean Common Market was permitted 'observer' status. Its intervention in negotiation of Lomé I had, of course, been more active than this.
86 The most explicit proof of this is the absence of an eastern Caribbean interest at the ambassadorial level. The ECCM were not represented and the issue of greatest significance to them, the banana protocol, was left to Jamaica, which did not press it. Source: interview.
87 Source: interviews.
88 See *Statement by the Council of ACP Ministers ... on the successor arrangement to the Lomé Convention*.
89 'Interview with D. Rainford, Chairman of the ACP Committee of Ambassadors', *The Courier*, No. 54, March–April 1979.
90 See the comments by Rainford in *The Courier*, No. 58, November 1979.
91 Comments by B. St John, *The Courier*, No. 58.
92 The complete text of Lomé II is reproduced in *The Courier*, No. 58. A useful commentary is found in Information, *Lomé II: Analysis chapter by chapter of the EEC-ACP Convention*, No. 194/X/80-EN, Brussels.
93 Source: interview.
94 Hewitt and Stevens in Stevens (ed.), *EEC and the Third World*, p. 50.
95 Figures calculated from *La CEE et la région des Caraibes Lomé I et Lomé II*. For the MDCs the total is from 52.5 mn ECU to 61.3 mn

ECU; for LDCs, 26.4 mn to 28.4 mn; and regional, 70 mn (including sums for Suriname and the Netherlands Antilles).

96 Island states are the Bahamas, Barbados, Dominica, Grenada, Jamaica, St Lucia and Trinidad and Tobago. Least developed are Dominica, Grenada and St Lucia. St Vincent has recently been accorded both 'island state' and 'least developed state' status.

97 *The Courier*, No. 58.

98 Figures calculated from *La CEE et la région des Caraibes Lomé I et Lomé II*. Individual targets are: Bahamas, 5 million ECU; Barbados, 9 million ECU; Guyana, nil; Jamaica, 10 million ECU; Trinidad and Tobago, 15 million ECU.

99 P. J. Patterson, addressing the Special Ministerial Conference on ACP Co-operation, Jamaica, July 1980, *The Courier*, No. 63, September–October 1980.

100 'The Road to Lomé: Some Thoughts on the Development of ACP-EEC Relations' (Memorandum prepared by J. O'Neil Lewis, Brussels, 9 September 1981, para. 162. O'Neil Lewis was Trinidad and Tobago's ambassador in Brussels from 1973 to 1982.

Vaughan A. Lewis

9 Commonwealth Caribbean relations with hemispheric middle powers

As is apparent from even casual observation, a wide gap separates the Commonwealth Caribbean states in material terms from many surrounding continental states of Central and South America. We can follow Spiegel in categorising Brazil and Mexico as middle powers, defined as 'those states whose level of power permits them to play only decidedly limited and selected roles in states and regions other than their own'.[1] Writing before the revolution in petroleum prices, he places Venezuela in the category of minor powers, although this event, which has so substantially increased Venezuelan financial resources, would seem to justify placing that country in the same category as the others. Spiegel also places Cuba in the category of minor powers, but in some respects the distinguishing criteria between minor and middle powers are rather blurred; differences in population, and thus of manpower resources, are, however, obviously important. In addition to material differences, it can plausibly be said that a substantial cultural gap also separates the two sets of states in terms of language and traditions; though it must be observed that this has not, historically, inhibited large popular migrations from some Commonwealth Caribbean countries, for example Jamaica, to continental locations involved in intense economic activity requiring surplus skills, such as Panama in the context of the Canal and Costa Rica to establish banana plantations.

In examining material differences, the most apparent is that of mere land area available to the various states. The surface area of the islands of the whole Caribbean area is approximately 238,000 sq km, smaller than that of either Mexico, Venezuela or Brazil. Guyana has a relatively large land area (214,000 sq km), but her population is miniscule (701,000) when compared with the continental states and even with some archipelago countries.[2]

The larger Caribbean islands bear some comparison in terms of *per capita* income, though not, of course, in terms of the actual sizes of gross national product. Brazil, Venezuela, Mexico, Jamaica, Barbados and Trinidad and Tobago all have *per capita* (GNP) incomes within the range of US$1,000–$3,000, thus placing them firmly within the ranks of middle-income countries.[3] The petroleum price explosion and discovery of new resources in Mexico, Venezuela and Trinidad have further distanced them from all other states in the region; and the size of the petroleum and natural gas industries of Mexico and Venezuela has distanced these states from Trinidad itself, in terms of disposable national income. In so far as financial capability is an important indicator of, or base for, the exercise of influence in inter-state relations, then their relative status has *prima facie* been increased – especially as many other states in the hemisphere have had difficulty in the face of the recent and continuing international economic crisis in maintaining their levels of disposable income.

Of course, the general underdevelopment and relative dependence of even these middle powers need to be emphasised, as conditioning their status in the global system and circumscribing their capacity for independent action *vis-à-vis* the dominant power. As one economic analyst of the region has written:

> ... Successive takeoffs have usually been terminated by balance of payments crises and/or political instability exacerbated by social tensions related to the preceding phase of inequitable economic growth and its denouement. In consequence, Latin America's power assertiveness has not risen commensurately with its economic advances ... the dependence of its growth on external stimuli to exports and on infusions of foreign capital and technology diminishing only modestly in the course of the past century. This failure to grow into an autonomous centre of international economic and political power is rooted in turn in the failure to achieve sustained broadbased economic growth and distributional equity.[4]

Nonetheless, in relation to the Caribbean and Central American subregion, the relative status of Mexico and Venezuela has undoubtedly risen, as oil, their 'plantation staple', has experienced a new upswing. The extent to which this status ascendancy has allowed them to assert or reassert claims to a legitimate role in the determination of political processes in the region, against the traditional and unilateral orientation of the United States, is more difficult to determine and is discussed later in this chapter.

Given the wide differences between the continental states and the Commonwealth Caribbean countries in terms of population and economic development, it is natural that these should be reflected in the relative sizes of the military and para-military (manpower) forces available to the respective states. While the military and para-military forces of Jamaica or Guyana do not (1978) figures exceed 15,000, the regular forces of Mexico number approximately 100,000, and those of Venezuela, 41,000. It has, for example, recently been observed that 'once logistics and training programmes are carried through ... [the Venezuelan] navy will become the most important in the Caribbean'.[5]

Finally, as middle powers in an hierarchical though partially decentralised contemporary international system, Venezuela, Mexico and Brazil have tended to be attributed roles in the partial international system (the hemispheric system) in which they participate with the dominant partner, the United States. Sharing the dominant systemic values about order and security that emanate from the United States, they have tended to be characterised by that country as having specialised responsibilities in the system – hence their designation, for example, as 'regional influentials' within it.[6] Yet, especially in respect to smaller states, they are aware of the dangers of being seen as purveyors and maintainers of US conceptions of order. Hence a continuing tension exists for them between being perceived as 'influentials' in the system on the one hand, and 'proxies' of the system on the other. Differing perceptions play some part in their capacity to exercise authoritative activities in the 'spheres of influence' which, partly unilaterally, they determine for themselves in the region.

Geopolitical orientations of the middle powers

The Commonwealth Caribbean countries constitute, as indicated above, a relatively small segment of the hemispheric system in terms of size, financial resources and security capabilities. Their relationship to the middle powers has traditionally been geopolitical – based on geographical proximity – rather than structural in the sense of exhibiting major economic linkages.[7] At the same time, however, Venezuela, Mexico and Brazil have been involved in other sets of geopolitical relations with (other) geographically proximate states, to which, in a hierarchy of diplomatic attention and awareness, greater importance has been attached.

Thus, while asserting periodically the importance of the Caribbean as her natural maritime outlet, Venezuela has traditionally exhibited a greater geopolitical concern with her immediate neighbours in north-western South America. On the other hand, 50% of her major export commodity, petroleum, is traded with the United States, which is also the source of nearly 50% of her imports (by value). Finally, as a founder member of the Organisation of Petroleum Exporting Countries, a major arena of her diplomacy is to be found in that sphere.[8]

The major geopolitical relationship for Mexico, on the other hand, derives from the fact of her 2,000 mile border with the United States. This relationship has been the basis of the Mexican concern with a diplomacy of non-intervention, while at the same time (as in the case of Canada) determining the nature of her trading relations. Approximately 57% (in value terms) of Mexico's imports derive from the US, while 62% of her exports go to that country (1976 data).[9] The country's discovery of new petroleum resources is increasing this concentration. Thus her economic and political diplomacy is, and will be, dominated by these relationships. In geopolitical terms, her next most salient relationships derive from her territorial contiguity with the Central American states, and the geographical proximity of Cuba to the Gulf of Mexico.[10]

The geographical size of Brazil has been the basis for her arrogating to herself a series of geopolitical concerns based historically, first on the sharing of a common value system about hemispheric security with the United States; and secondly, on the notion of absolute security at her frontiers. As is widely appreciated, Brazil, perceiving Argentina as her most assertive potential competitor for status on the continent, engages continually in a competitive and co-operative relationship with that country, using the intervening buffer states (Bolivia, Paraguay, Uruguay) as stakes in that relationship and, therefore, as arenas for competition for influence. Venezuelan claims to substantial territory of Guyana, with which Brazil shares a border, have now induced a second sub-regional zone of competition in which Brazil has become involved. Nonetheless, this zone clearly concedes precedence in the hierarchy of Brazilian concern to the Bolivia–Paraguay–Uruguay zone.[11]

In large measure, the styles and content of these orientations, and indeed their competitive character, have been traditional and non-ideological, focusing mainly on the search for material accumulation and for spheres of influence for maintaining geo-strategic security.[12] In that context, the Commonwealth Caribbean states have come low in the middle powers' hierarchy of concern – largely because they are relatively new entrants into the hemispheric interaction system and, protected hitherto by the United Kingdom, did not constitute a 'vacuum'. The previously independent Greater Antilles (Cuba, Dominican Republic, Haiti) had traditionally fallen under the security surveillance of the United States. Thus the case of the Brazilian extension into the Dominican Republic (1965) under the umbrella of the Organisation of American States is properly seen as a means of the new military regime signalling reinforced allegiance to United States Cold War conceptions, within the context of the Brazilian notion of 'ideological frontiers'.[13] And in much the same way, the developing relationship between the Quadros regime and the Cuban revolutionary regime at the beginning of the 1960s is more indicative of an attempted reorganisation of US–Brazilian relations than of a major Brazilian reorientation towards the Caribbean.

The middle powers' relationships with the newly sovereign Commonwealth Caribbean states take place, then, in the context of a complex of hemispheric interactions, at differing levels of action. These levels of action are seen as being based on particular regional centres of power,[14] essentially what we have been calling middle powers, with Brazil seen as the power centre, having the most substantial potential for upward mobility into the category of secondary powers.[15] The complex of interactions involves Brazil, Argentina, Venezuela and increasingly Mexico, as that country increases its financial weight in the system. The levels of interaction, whether between Brazil and Argentina, Brazil and Venezuela, Brazil and Mexico or Mexico and Venezuela, extend over a wide range of issue-areas encompassing security, economic relations and the area of nuclear technology. They are unstable in the sense that they extend from coherent to antagonistic and disintegrating, depending on individual countries' perceptions of how the gains from interaction are likely to enhance the relative status of one or other state.

One implication of this mode of interaction for smaller powers or groupings is that, given the differences in capabilities between themselves and the middle powers, they are perceived as stakes, and are likely to find themselves incapable of sustaining the allegiance or protection of such powers consistently over time. Even within the restricted international sphere of regional politics, the middle powers are unlikely to be consistent patrons. The instability of the interaction systems inhibits the development of sufficient hegemony by any single power.

A second implication is that the styles of diplomatic activity among the centres of power on the one hand, and between any particular centre and the smaller power or powers on the other, are likely to differ. In the first case, diplomatic activity takes the style predominantly of bargaining relationships. In the second case, it involves a combination of bargaining and command politics on the part of the middle power. The relationship between Mexico and Jamaica in respect of the Javamex alumina smelter is a striking recent instance of this. Mexico, a major partner in this venture, seen as an important project by the Jamaican government, peremptorily withdrew after repeated assurances as to her continued participation.

And a third implication, historically familiar in international relations, is that such middle powers, with some sense of capacity for partially autonomous activity in global relations, are likely to support smaller sub-regional systems for reasons extraneous to the objectives of the smaller powers. One analyst argues, for example, that Venezuela's entry into the Andean Common Market was 'in part ... directed toward counteracting Brazilian imperialism in Latin America', the then President Calvani taking the view that 'the growing economic and political influence of Brazil in the Caribbean as well as the Andean area was a major concern'.[16]

At the level of the international system, the orientations of these middle powers, especially since the period of *détente*, has been to diversify their economic and diplomatic relations away from the United States. This has taken different forms in particular countries: the 'pragmatism' of Brazilian diplomacy, directed at enhancing the economic growth and status of the country; the 'diplomacy of projection' of Venezuela, concerned to establish coherent relations with non-hemispheric middle powers: or the diplomacy of reform of the international economic order of Mexico.

For the Commonwealth Caribbean states the point of significance of these diplomatic orientations is that none of the states undertaking them perceives itself capable of sustaining alliances of smaller powers involving the distribution of gains from these endeavours. For this, the smaller powers would still have to seek relationships beyond the diplomatic and other institutional arenas dominated by middle powers. Within this broad understanding, then, of the orientations and styles of, and limitations on, middle powers in the hemisphere, we turn now to identify the specific interests that they pursue in the Caribbean sub-region.

The interests of Venezuela[17]

Venezuelan interests in the area can be defined in terms of two sets of factors. The first of these relates to her geographical contiguity to Guyana on the mainland, and her geographical proximity to Trinidad and Tobago, which is located just off the northern coast of Venezuela. With the imminent independence of the then British Guiana, Venezuela revived claim to virtually two-thirds of the territory of that country. The so-called Geneva Settlement, committing the two countries to peaceful means of resolving the dispute, was followed by the Protocol of Port-of-Spain in 1970, 'freezing' the dispute for a period of twelve years. This followed a period of intense diplomatic activity by the Guyanese government involving a degree of internationalisation of the dispute and the search for diplomatic support at the United Nations and within the Third World.[18]

The government of Guyana, in the course of its diplomatic activity, was able to obtain the support of Brazil to the extent that that country's government was willing to assert as its position the maintenance of existing boundaries, and the muting of its own lesser territorial claim against Guyana, as long as there was no aggressive territorial move by Venezuela. In fact a triangular relationship between Guyana, Venezuela and Brazil has developed on this issue-area, in which one of Guyana's fundamental objectives is to conduct internal and external policy in such a manner as not to alienate the support of Brazil. Venezuela has, in turn, become cognisant of the fact that the assertion of her interest through the territorial claim is constrained by the protective stance of Brazil.

The maritime dispute between Venezuela and Trinidad is a less severe one, though the difficulty in resolving it has led to a degree of diplomatic hostility towards Venezuela on the part of the Trinidad government, and claims by the latter of 'Venezuelan imperialism' in the Caribbean.[19] This dispute is indicative of the second set of factors through which Venezuela asserts a national interest. Venezuela claims to have a general diplomatic and strategic interest in the evolution of relationships among the countries of the Caribbean Sea. As her representatives have often claimed (partly in the face of opposition to her assertion of an interest here), the Venezuelan coastline on the Caribbean is the longest of any Caribbean country (3,000 km).[20] It is pointed out that the important commercial and industrial centres of Venezuela (Caracas, Maracaibo) face the Caribbean Sea, and that outlets on the Caribbean Sea are of prime significance for the exports of oil and oil products which constitute the major source of the country's foreign exchange. 'Venezuelan history,' President Carlos Andres claimed at the third UN Conference on the Law of the Sea, 'had developed along the Caribbean and largely under its influence', though on the other hand 'the country had never applied a policy towards the sea'.[21]

In applying a policy so as to consolidate her claimed interests in the sea, Venezuela has, in recent years, initiated a series of boundary delimitation and fisheries agreements with other Caribbean and Caribbean-related states. Such agreements have involved a reassertion of Venezuelan possession of the minute Aves Island in the northern Caribbean. Its importance derives from the growing consensus at the third UN Law of the Sea Conference that islands, properly defined, might be entitled to their own law of the sea regime. Such a regime as relating to Aves Island would give Venezuela an even more integral presence in the Caribbean, and project her strategic interests even further.

Venezuela's asserted interests have led to the elaboration of a variety of economic aid relationships with Commonwealth Caribbean countries, both dependent and independent, especially in the Windward and Leeward Islands. In 1975, for example, within a four-month period no less than five Commonwealth Caribbean heads of state, those of Jamaica, Guyana, Antigua, Grenada and St Kitts-Nevis, visited Caracas with a view to signing economic co-operation agreements. Since the rise in petroleum prices this policy has also manifested itself through the establishment of concessional arrangements for the provision of petroleum to Central American and then to Caribbean countries, participation in the Caribbean Development Bank, and the more recent joint Mexican–Venezuelan facility for new concessional arrangements for petroleum to a variety of states in the Caribbean Basin (the San José facility).[22]

There has been a suggestion, here, of the Caribbean as a Venezuelan 'sphere of influence' in the traditional sense, which has met with some degree of hostility from Trinidad and Tobago, also a recipient of oil revenues and concerned to elaborate her own arrangements for financial assistance towards the sustenance of what that government sees as the Caribbean Community identity. Venezuelan assertions of the need to protect 'democratic values' in the area have also been partially responsible for developing ideological and diplomatic competition with Cuba.[23]

The interests of Mexico

Mexico has traditionally demonstrated little concern with the Commonwealth Caribbean territories. Ex-president Lopez Portillo has proferred the view that

> ... unlike the continental mainland, where the principal nations of Europe established their culture over widespread areas, those [Caribbean] islands became the sites of isolated enclaves ... the emerging nations of the Caribbean constitute a new geopolitical configuration. The lack of meaningful contacts with their nearest neighbours – the inevitable result of colonial organisation – is gradually being replaced by a conscious search for regional exchanges.[24]

Within the Caribbean, Mexico's major area of concern, in the postwar period, has been Cuba, which points into the mouth of the Gulf of Mexico, and thus at her gateway to the Atlantic. But even in respect of the revolutionary regime in Cuba, Mexican policy has mainly been aimed at inhibiting American tendencies to draw the Latin American states into justification of overt intervention in that country. More positively, she has sought to encourage the Cuban government towards concentrating on the primacy of material (economic and financial) relationships with the hemisphere over ideological ones.[25]

In a chronological sense, Mexico's central concern with the Commonwealth Caribbean is, in fact, an extension of her interest in Central America. Her efforts have been directed at inhibiting, through diplomatic means, any attempt by Guatemala to make real by physical means her territorial claim to Belize. In pursuit of this aim, she has muted her own territorial claim to Belizean territory. And, at the urging of Commonwealth Caribbean states, she has led other Central American states and Venezuela against the trend of their historical regional diplomacy in agreeing that Belize should become a sovereign state, with necessary territorial adjustments being subject to normal international negotiation. Mexico thus maintains a general strategic and security interest in the evolution of Guatemala–Belize relations, which continues even now that Belize has become a sovereign state.[26]

More recently Mexico has had a co-operative-competitive relationship with Venezuela in respect to Central America. The more co-operative aspects of this (the oil facility and technical assistance to particular countries) have been extended to the Caribbean countries. It is aimed partly at demonstrating the increasingly insistent Mexican view that economic and technical assistance, rather than security assistance, constitutes the optimal mechanism for developing sociopolitical stability in the Caribbean Basin, and that, in Lopez Portillo's words, 'collaboration between Venezuela and Mexico is vital to achieve the stability of the 27 nations in the Caribbean Basin'.[27] This view has also conditioned the nature and extent of Mexico's participation within President Reagan's 'Caribbean Basin Initiative'.

Mexico was minimally responsive to Jamaican political and financial diplomacy during the 1970s, at a period when there was a certain coincidence of aims between the two countries in respect of reform of the international economic order. In that context, too, the CARICOM-Mexico Joint Commission was established; and Mexico now seeks to multilateralise her assistance to the Caribbean Community countries by seeking membership as a contributing member to the Caribbean Development Bank.[28]

In general, therefore, and contrary to the position of Venezuela, Mexico's relatively restrained activity in the Caribbean area suggests that there has been a limited identification of interests in the sub-region. Mexican diplomacy towards the South-Central American zone has tended to be subject to the pre-eminence of her relationship with the United States, which is based fundamentally on their geographical contiguity. Within that framework, however, her new oil-derived financial strength seems capable of allowing her to exhibit a greater assertiveness in a role which she likes to see as appropriate for herself: that of diplomatic protector of the rights of small countries versus the more powerful ones. What was once largely a diplomatic and reactive stance can now be given some positive, more material, content.

The interests of Brazil

These are perhaps the least formally operationalised of the three middle powers,[29] though Brazil has perhaps the most experience of dealing with small powers on her periphery. Her most direct intervention in the archipelago has been, as indicated above, through security assistance to the US–OAS forces during the intervention in the Dominican Republic in 1965. This mirrored the new Brazilian military regime's anti-Communist, anti-Cuban stance at home. Given the Brazilian regime's concern to ensure the impenetrability of the country's frontiers by 'alien' forces, its fundamental attitude towards the Caribbean would appear to be premised on geo-strategic views.

This would seem to be the basis of its approach to Guyana, with whom Brazil shares a boundary. While the cultural similarities between Brazil and her small South American neighbours are likely to support a greater degree of direct intervention, the general principle that she applies to them seems to apply to Guyana also. This is, that the peripheral small state should not seek alliances with forces or entities known to be ideologically or geo-strategically hostile to Brazil. Also implied in this stance is the suggestion that a peripheral small power should not attempt to sustain a *domestic* regime which can be maintained only by substantial assistance from states known to be hostile to Brazil.

We have already alluded to the existence of a triangular diplomatic relationship between Venezuela, Brazil and Guyana, deriving from the Venezuelan–Guyana territorial dispute. This relationship is likely to be joined by the small and weak state of Suriname, which has also asserted a claim to Guyanese territory. Within this system, both Brazil and Venezuela seek to influence the smaller entities through forms of economic and technical assistance. For Guyana is posed the question of whether, over the long term, this system of relations will gain greater salience for her sovereignty and diplomacy than the CARICOM system.

Middle power interests, Cuba and the Caribbean

All the middle powers have developed a strong awareness of what they perceive as the current and potential influence of Cuba in the Commonwealth Caribbean region. The decade of the 1970s in particular was characterised by a dual policy on the part of these states and others in Latin America towards that country (Brazilian diplomacy has been somewhat more rigid than others towards Cuba, and what follows applies less to her government). On the one hand, they have sought to induce her re-entry into the inter-American system, and to legitimise her participation in the developing diplomatic relations of the Caribbean and Central America. On the other hand, they have sought to play a role in the determination of the nature and limits of the influence that Cuba might exercise in the area. This applies in particular to Venezuelan diplomacy.

There is a widespread awareness that, partly on the basis of her security and military aid relationship with the Soviet Union, Cuban interests and diplomatic-political activities are more extensive over the globe than her physical and/or economic size would suggest. On the other hand, Cuba's foreign relations can, from her perspective, be said to be based on a global ideology of anti-imperialism and national liberation (defined to include both anti-colonialism and internal liberation), underpinned by the philosophy and praxis of Marxism-Leninism.[30] Within this perspective, she asserts an identity of entitlement to full participation in the Latin American and Caribbean partial international system, not accepting any special legitimacy of, for example, Venezuela to assert limitations on her activity.

Cuba's relations with the middle powers tend, therefore, to vary in intensity and distance according to issue. This is the consequence of Cuba's revolutionary assertiveness and the middle powers' periodic rejection of this, depending on their assessments of the implications of such assertiveness for their domestic systems, their relative status in the region, and their particular relationship with a United States normally hostile to Cuba. Their relationship with Cuba tends, as a result, to be competitive.

This competitiveness is reinforced by a revived concern with the Caribbean as an important area of passage for petroleum and petroleum products in a period of general resource scarcity. It is to be placed in the context of a general Western concern with the possible constraints, at the global level, on the availability of scarce resources located in unstable or undependable regions:

> For what is emerging is nothing less than a remarkable new strategic map. The practical effects are to resurrect the importance of geography and resources as a factor in military thinking, and to make us more sensitive to the geo-strategic perspectives of regional powers.[31]

The significance of these types of middle-power relationships with Cuba is, for our purposes, that the perceived extent of Cuban activity in the area partially defines – particularly in the case of Venezuela – the context in which the middle powers perceive the activities of the Commonwealth Caribbean countries. So Venezuela's attitude to them, for example, takes on the aura of being a function of Cuban–Venezuelan activity and systematic competition in the sub-region. The particular dynamics of the US–Cuba conflict become a constraining or permissive input into the definition of the Venezuelan attitude. And, in the case of the Venezuela–Guyana dispute, there would appear to be an element of manipulation of the perception of the salience of Cuba's presence and activity in Guyana.

In summary, then, on the basis of their perceptions of Cuba's role, the level of and receptivity to it on the part of other regional countries, and their structural and strategic interests in the area, we can define the contemporary behaviour of the middle powers in the following way:

Venezuela has adopted an attitude of 'assertive interventionism', claiming a legitimacy borne of geopolitical location, for active participation in the nature of the development of the Caribbean as a diplomatic and strategic arena. Her increased financial resources provide additional capabilities for operationalising such a role.

Mexico can be said to have undertaken a role of 'reactive or protective diplomacy', still more concentrated in the western as against the eastern Caribbean. This represents a deviation from her tradition of non-intervention, and is appropriate to her perception that her present resources can permit her, selectively, to become a countervailing force and buffer *vis-à-vis* American attempts to continue to define unilaterally the nature of, and constraints on, the regional sub-system.

Brazil continues to assert a role essentially of command politics *vis-à-vis* small powers, determined by her traditional geopolitical view of Latin American international relations. Within this context, she is willing to offer peripheral states participation in her economic development process, though the experience of Commonwealth Caribbean states in one crucial issue-area, fisheries, has not been particularly beneficial.

Perspectives and interests of Commonwealth Caribbean countries

It might be said that the immediate post-independence history of the larger Commonwealth Caribbean states suggests an experience of recalcitrance and perceived difficulty in attaining objectives *vis-à-vis* neighbouring states of the South American continent. Speaking in the first parliamentary debate on foreign affairs in an independent Trinidad and Tobago in 1963, Prime Minister Eric Williams, while noting that Trinidad had 'developed close contact in the United Nations' with the Latin American group, went on subsequently, in reference to the possibility of joining the Organisation of American States, to assert that 'there has been a feeling in our direction of a certain resentment that our rights as a member of the American family are not recognised, and that we have to depend upon what ultimately appears to be something of grace instead of... as of right'.[32] In the same speech he made reference to the problems posed by territorial disputes for countries (he was clearly referring to Guyana) wishing to enter the Organisation of American States and its allied institutions. Perceptions softened in subsequent years, but the decision by Venezuela to 'de-freeze' the Protocol of Port-of-Spain appears to have revived this imagery. Moreover, the speed of Venezuelan emplacement of a diplomatic presence throughout the Caribbean, allied to promises of substantial economic and technical assistance, led Williams in 1979 to remark on the 'relegation of the Caribbean to the sphere of influence of Latin America'.[33] Analyses of regional politics have also attempted to discern whether the growing influence of Venezuela might damage the performance of CARICOM.[34]

Some of the difficulties alluded to by Eric Williams in his various observations on Venezuelan influence reflect in part the very proximity of Trinidad to Venezuela, and the difficulties arising therefrom. Williams had also periodically alluded to the necessity to recognise, and sustain, the Caribbean archipelago countries and the mainland Guyana territories as having a cultural and ethnic identity different from that of the rimland states surrounding them.[35] With additional Venezuelan financial resources after 1973, the question began to arise as to whether Trinidad or Venezuela would have the greater influence either in sustaining or breaking down this claimed particular identity. It might be asked also of those who hold the view of the 'separate identity' whether Venezuela can claim any legitimacy for participation in and direction of the Caribbean arena as long as she holds a significant claim against the territorial integrity of Guyana; and whether her assertion of this claim might not be deemed more threatening than the assumed threat of expansion of Cuban influence in that zone.

The debate concerning the relative strengths and consequences of Venezuelan as against Cuban influence has, however, also reflected differences within the Commonwealth Caribbean itself as to (a) the relevant scope of the Caribbean Community as a viable diplomatic actor, and (b) differing perceptions among constituent states, especially in an era of financial and resource scarcity, as to where such resources might be derived and the nature of the relationships which needed to be made to ensure their derivation. These differences were manifested in a number of ways during the 1970s.

In the first place, there was disagreement within the Commonwealth

Caribbean region about the proper scope of regional relations. The antagonism felt by the Trinidadian government about the role being played by Venezuela was directed just as much against Jamaica as against Venezuela itself. During the 1970s the Jamaican government developed a view of the Caribbean which diverged from the 'particular identity' concept of Trinidad. Speaking just over a year after coming to office, the Jamaican Prime Minister Michael Manley remarked on the fact that Jamica, as a relatively small island, was 'surrounded by a continent of Latin American peoples'. And in his work *The Politics of Change* he envisages a system of 'economic regionalism' which would encompass the Central American states, the Basin countries of Colombia and Venezuela, and the north-eastern South American countries of the three Guyanas and Brazil.[36]

Secondly, there developed in the 1970s a certain convergence (not identity) of diplomatic view, within the context of global *détente*, on appropriate policy towards the Third World, non-alignment and international economic reform between two major Commonwealth Caribbean states (Guyana and Jamaica) and two of the proximate middle powers (Venezuela through the policy of Carlos Andres Perez, and Mexico through the policy of Echeverría). The Latin American partial international system has not had, with the exception of the experience of Cuba, a tradition in which minor states within it sought to undertake an autonomous diplomacy in respect to issue-areas that are global in scope. Yet, in the final analysis, given the differences in economic weight and therefore in diplomatic leverage, it was not possible for the two small Commonwealth Caribbean states to form an effective hemispheric alliance with the two middle powers on terms that might demonstrably provide benefits appropriate to their expenditure of resources on the negotiating process. There is here, also, a difference in perspective about the scope of the time period in which benefits need to be derived. The subsequent formation of the Latin American Economic System (SELA) did not, nor could it, negate this problem. For SELA does not function as a collective substitute, but as an additional negotiating resource. The Commonwealth Caribbean states' foray into diverse international relationships, and the search for supporting relationships among the middle powers, did not therefore, given the discrepancies of perspective and resources, provide any incremental support for the development of the Caribbean Community system as an effective diplomatic force.

A third area of Commonwealth Caribbean dissension has been the development in the late 1970s of political relationships based on ideological allegiances. Countries oriented to various forms of socialism have sought to consolidate state-to-state relations with the ideological cement of party-to-party relationships. This, in turn, revived the traditions of international alliances between those of liberal or Christian Democratic persuasion. Each grouping of parties-states has, in turn, sought the support of European sources from which the ideologies derive. And, in a period of relative American abstinence and partial discredit of American-sponsored development strategies, political factions in various European countries have, to some extent, been able to exert influence on local state foreign policy preferences. In return, local state decision-makers have been able to draw institutional and material support from the European sources.[37] The main question here is whether such institutional and ideological linkages can withstand sustained United States opposition within a global Western context, and in which the leverage of the Caribbean sector is minimal.

Conclusion

This chapter has argued that for most of the 1960s and 1970s the relationships between the middle powers – Venezuela, Mexico and Brazil – and the Commonwealth Caribbean states can be classed broadly into the following categories: (a) economic aid and trade, with aid the dominant component, (b) geopolitical, deriving from geographical proximity, and (c) institutional, deriving from the efforts of Latin American and Caribbean states as a whole to seek forms of collective economic security.

Within the sphere of economic aid relationships, the most important of the powers involved has been Venezuela, which has sought to grant its assistance through a mixture of bilateralism and multilateralism. Bilateralism, concentrated largely on the smaller Commonwealth Caribbean countries, was meant to ensure that, with the gradual withdrawal of the British, no influence vacuum might be permitted to develop. In addition, Venezuela seems to have sought to ensure that her own aid relationships with these states were sufficiently continuous and predictable to guarantee that her own influence might be a necessary (though not the only) one in their decision-making about the types of regional and international relations they might embark upon after independence. In a number of cases Venezuela's aid to these territories has been in train when they have been still, formally, dependents of the

United Kingdom. This orientation on the part of Venezuela was first apparent after 1970, the year in which political instability developed in the largest of the eastern Caribbean states, Trinidad and Tobago, the government of which appears to have appealed to Venezuela for substantial assistance to quell the uprising. This event will certainly have afforded Venezuela a feeling of legitimacy for, if not an interventionist, certainly an 'overseer' role in the region.

Secondly, the early 1970s were a period in which Venezuela's continental neighbour, Guyana, was itself attempting to consolidate its diversification of relations towards the world socialist system. This involved first the normalisation and then the institutionalisation of relations with Cuba, and secondly the reorganisation of its political and economic institutions along Marxist-Leninist lines in order to make them more appropriate to deepened relationships with the socialist bloc. These innovations were taking place virtually simultaneously with Guyana's attempt to reinvigorate the movement for political integration in the eastern Caribbean by seeking, through the Grenada Declaration of 1971, to establish a unitary political state with the countries of the area. Venezuela will have seen it as its task to combat such Guyanese influence, through *inter alia* the use of economic instruments.

Finally, as the Commonwealth Caribbean governments sought to normalise their relations with Cuba, and Cuba in turn sought to establish a presence in the Commonwealth Caribbean, Venezuela will have gained an enhanced sense of increasing competition for influence in the sub-region. Within the area, this competitive influence-seeking involved Venezuela, Trinidad and Tobago and Cuba, the former two having had their capabilities for exercising influence increased by the consequences of the petroleum price revolution. Venezuela's multilateralising of a portion of her economic aid through the Caribbean Development Bank will have been seen as a means of further legitimising her Caribbean presence.

All these factors represented an attempt by Venezuela to widen the boundaries of the sub-region, and in effect to increase the number of legitimate participants within it. In this she was assisted by Jamaica (in the second half of the 1970s), whose government sought to bring into effect a wider conception of a viable Caribbean system. This, as we have suggested earlier, would, for Jamaica, have served the purpose of bringing more extensive resources into the system. The government of Trinidad saw this, however, as weakening the 'integrity' of the original system. This attempt to give the regional relations in which the

Commonwealth Caribbean states were involved a wider base had, therefore, the paradoxical result of decreasing the coherence of the CARICOM system itself.

At the same time, relations between Jamaica on the one hand and Venezuela and Mexico on the other had, by the end of the 1970s, suggested that Jamaica had not autonomously developed sufficient 'weight' to ensure that they would be conducted with predictability, or that in a situation of *de facto* asymmetry these middle powers would not resort to the unilateralist orientation of command politics. Trinidad, however, on the basis of her oil wealth, could seek to establish trade relations with Brazil, based on market principles of comparative advantage.

At the level of geopolitical relations, the central question would appear to be whether relationships with larger states on the periphery of the Commonwealth Caribbean will, over time, gain greater salience for the sub-regional states involved, with one of two effects: (a) forcing a widening of the CARICOM system institutionally to accommodate these relations with Venezuela, Brazil and Guatemala through her relations with Belize, as functional members of the system; or (b) a 'tearing off' of the states with increasingly coherent relations with the peripheral middle powers, and their absorption into new systems of relations. This applies most forcefully to Guyana and Belize.

Finally, akin to the sphere of geopolitical relations, there lies an institutional question which derives from the perspective of at least some of the middle powers that the states of the Commonwealth Caribbean sub-region are essentially 'objects' of international relations, too weak individually to resist manipulation but with insufficient resources to organise a collective system incapable of undesired penetration. This view can be read into the diplomatic orientations of Venezuela and Brazil in particular. It implies a level of middle-power interaction, sometimes competitive but largely co-operative, in circumstances where the influence of Cuba appears to be becoming pervasive. This level of interaction would imply role responsibilities for particular middle powers, with respect to particular units or groupings within the CARICOM area. On the other hand, this geopolitical perspective would not see CARICOM as a functional unit, but the states of the wider Caribbean archipelago and/or Basin as the more relevant one. It is doubtful whether a concerted attempt to use resources and capabilities to maintain stability over this wider sphere would leave the CARICOM system as an identifiable one in future international relations.

Notes

1. Steven L. Spiegel, *Dominance and Diversity: The International Hierarchy*, Boston, 1972, p. 99.
2. United Nations, *Demographic Yearbook 1978*.
3. International Bank for Reconstruction and Development, *World Bank Atlas 1978*.
4. David Felix, 'Latin American power: take-off or plus c'est la même chose?', *Studies in Comparative International Development*, XII, 1, 1977, pp. 59–85.
5. Robert L. Scheina, 'Regional reviews: Latin American navies' in *U.S. Naval Institute Proceedings*, March 1981, p. 23. The data on military force are taken from Institute of Strategic Studies, *The Military Balance 1979–80*, London, 1979.
6. This phrase was much used by Zbigniew Brzezinski, formerly President Carter's National Security Adviser.
7. See R. Preiswerk, 'The Relevance of Latin America to the foreign policy of Commonwealth Caribbean states', *Journal of Inter-American Studies*, 11, 2, 1969, pp. 245–71.
8. For an analysis of Venezuelan foreign policy see R. D. Bond (ed.), *Contemporary Venezuela and its Role in International Affairs*, New York, 1977.
9. Data on the structure of trade are taken from United Nations, *Statistical Yearbook, 1978*.
10. For an analysis of Mexican foreign policy see M. Ojeda, *Alcances y límites de la política de México*, Mexico City, 1976.
11. For an analysis of Brazilian foreign policy see W. Perry, *Contemporary Brazilian Foreign Policy: The International Strategy of an Emerging Power*, London, 1976.
12. See the heavily traditional geopolitical analysis of some of these relationships by Lewis A. Tambs, 'Geopolitical factors in Latin America' in Norman A. Bailey (ed.), *Latin America: Politics, Economics and Hemispheric Security*, New York, 1969, and 'The changing geopolitical balance of South America', *Journal of Social and Political Studies*, 4, 1, 1979, pp. 17–35.
13. See in general F. Parkinson, *Latin America, the Cold War and the World Powers 1945–73*, London, 1974.
14. Wolf Grabendorff, 'Perspectivos y polos de desarrollo en América Latina', *Nueva Sociedad*, 46, 1980, p. 43.
15. For one perspective on this, see H. Jaguaribe, 'El Brasil y América Latina', *Estudios Internationales*, VIII, 29, 1975. And for some qualifications to Brazil's capacity for upward mobility in the short term see Wayne Selcher, *Brazil in the Global Power System*, Occasional Paper Series, No. 11, Centre for Brazilian Studies, SAIS, Johns Hopkins University, No. 5, 1979.

16 Donald L. Herman, *Christian Democracy in Venezuela*, Chapel Hill, 1980, pp. 179–83.
17 This discussion of Venezuelan interests, and the following one on Mexico, are in part drawn from Vaughan Lewis, 'Geopolitical realities in the Caribbean' in Elizabeth Thomas-Hope (ed.), *Caribbean Regional Identity*, forthcoming.
18 See Robert B. Manley, *Guyana Emergent: The Post-independence Struggle for Nondependent Development*, Boston, 1979, pp. 41–54.
19 See Eric Williams, 'The Threat to the Caribbean Community', Speech to a Special Convention of the People's National Movement, 15 June 1975; and more generally Henry S. Gill, 'Conflict in Trinidad and Tobago's relations with Venezuela' in L. F. Manigat (ed.), *The Caribbean Yearbook of International Relations, 1975*, Leyden, 1976.
20 See Vaughan Lewis, 'The Interests of the Caribbean Countries and the Law of the Sea Negotiations', Paper delivered to the Conference on Caribbean Maritime Issues, Florida International University, Miami, 12 April 1981.
21 Address delivered to the Third UN Law of the Sea Conference by the President of Venezuela, Carlos Andres Perez, Caracas, 20 June 1974. For a wider discussion see Demetrio Boersner, 'The policy of Venezuela towards the Caribbean' in L. F. Manigat (ed.), *The Caribbean Yearbook of International Relations, 1975*, Leyden, 1976.
22 For a description see *Bank of London and South America Journal*, 14, 4, 1980, p. 233; also *Latin America Weekly Report*, 10 April 1981, p. 12.
23 See for recent general discussions D. Boersner, 'La politique extérieure du Vénézuela: évolution et perspectives', *Relations Internationales*, 23, 1980, pp. 267–87; and F. Barthelemy-Febrer, 'L'offensive vénézuélienne dans les Caraibes: continuité et changements', *Amérique Latine*, 4, 1980, pp. 19–27. See also a recent Venezuelan newspaper editorial stating, '... el Caribe es el mare nostrum, el Mediterráneo natural de Venezuela', *El Diario de Caracas*, 1 February 1981.
24 Address delivered by the President of Mexico, Jose Lopez Portillo, at the dinner in honour of Mr Lynden Pindling, Prime Minister of the Commonwealth of the Bahamas, 23 February 1981.
25 See Olga Pellicier de Brody, *México y la revolución cubana*, Mexico City, 1972.
26 See on Mexico's view *Up-date: United States – Canadian/Mexican Relations*, Hearing before the Subcommittee on Inter-American Affairs of the Committee on Foreign Affairs, House of Representatives, Ninety-sixth Congress, 17 and 26 June 1980, Washington, DC, pp. 55 and 60.
27 'Mexico and Venezuela plan to counter outside intervention in Caribbean', *New York Times*, 9 April 1981.
28 Venezuela and Colombia contributed as members 13·2% of the CDB's total resources as of December 1979. CDB, *Annual Report 1979*, Barbados, 1980.
29 For an historical review that suggests this limited role see Cleantho DePaivo Leithe, 'Brasil y el Caribe', *Revista Argentina de Relaciones Internationales*, 15, 1979, pp. 53–64.

30 For an analysis of Cuban foreign policy see Cole Blasier and Carmelo Mesa-Lago (eds.), *Cuba in the World*, Pittsburgh, 1979.
31 Geoffrey Kemp, 'The new strategic map', *Survival*, 19, 2, 1977, p. 52.
32 *The Foreign Relations of Trinidad and Tobago*, 6 December 1963, Port-of-Spain, 1963, pp. 5 and 15.
33 Government of Trinidad and Tobago, *White Paper on CARICOM*, n.d., c. April 1979, p. 9.
34 See Anthony Payne, *The Politics of the Caribbean Community 1961–79: Regional Integration amongst New States*, Manchester, 1980, pp. 212–17.
35 Williams was contemptuous of the idea of a Caribbean Basin and in May 1975 succeeded in getting the Economic Commission for Latin America to create within its structure a Caribbean Development and Co-operation Committee, designed to preserve the separate identity of the Caribbean. See Paul Sutton (ed.), *Forged from the Love of Liberty: Selected Speeches of Dr Eric Williams*, Port-of-Spain, 1981, pp. 400–3.
36 Michael Manley, *The Politics of Change: A Jamaican Testament*, London, 1974. For a discussion of the Jamaican view of the region see Vaughan Lewis, 'The Commonwealth Caribbean' in C. Clapham (ed.), *Foreign Policy-making in Developing Countries*, London, 1976.
37 See Latin America Bureau, *The European Challenge: Europe's New Role in Latin America*, London, 1982.

Denis Benn

10 The Commonwealth Caribbean and the New International Economic Order[1]

Commonwealth Caribbean states, particularly Guyana and Jamaica, have played an active part in the discussions on the establishment of the New International Economic Order (NIEO) which have tended to dominate the agenda of international economic conferences during the past decade. This is not surprising, since Commonwealth Caribbean countries, like other Third World[2] countries, suffer a number of economic disadvantages resulting from the operation of the existing international economic system and therefore stand to gain from the establishment of a more rational and equitable order. This chapter examines the role of Commonwealth Caribbean states in the debate on the NIEO and the relevance of changes in the existing international economic system for their development. In the process an attempt will be made to identify the elements of what may be described as a Caribbean strategy for effecting fundamental changes in the present system.

The demand for a New International Economic Order

For some time there had been widespread recognition, at least on the part of Third World countries, that the structure of the existing world economic order perpetuated patterns of dependence between the developed and the developing countries and thus inhibited the development of the latter. This view derived from the argument that the present world economic order had grown out of a history of colonisation and imperialist domination and that its essential features had been consolidated at a time when most of the developing countries were appendages of the developed world. There was a conviction on the part of the developing countries, therefore, that this order which had emerged in the heyday of colonialism and which had been further consolidated in the immediate post-war period reflected the interests and preoccupations of the developed countries.

However, although the understanding of the problem was not new – it was certainly appreciated at the time of the first UNCTAD in 1964 – the inequities of the present system and the need for radical and fundamental changes in its basic structure were accentuated by changes in the world economy during the 1970s, beginning with the 1971 Smithsonian 'devaluation' of the American dollar, which signalled the virtual collapse of the Bretton Woods system, and culminating in the so-called 'international economic crisis', following the decision of the OPEC countries in 1973 to increase the price of oil and the consequent transfer of significant economic resources to certain sections of the developing world. These difficulties were compounded by serious inflationary trends, resulting in significant increases in the price of manufactured products, fertilisers and essential food supplies and also by marked signs of economic recession in many parts of the developed world.

What was also significant was the fact that the conventional economic wisdom, based largely on the Keynesian economic principles which had sustained the post-war international economic order, no longer seemed capable of explaining new phenomena such as 'stagflation', defined as the simultaneous existence of spiralling inflation with high levels of unemployment. The truth is that the post-war international economy had never experienced such a rapid succession of changes – all of them interrelated and mutually reinforcing in their impact. It proved virtually impossible in the short term to completely understand and assess their economic implications, much less control economic activity adequately through monetary policy and fiscal measures in a period of crisis and rapid economic change. Moreover, the recycling of petro-dollars, the rise in interest rates and the increasing debt of Third World countries presented new and complex challenges for effective economic management on a global scale.

Needless to say, the impact of these developments on the economies of the non-oil-producing Third World countries was, with few exceptions, disastrous. Not only did they result in a significant increase in the value of the imports of these countries relative to their export earnings, thereby adversely affecting their terms of trade, but there was also a significant increase in their external debt. It was estimated, for example, that the balance of payments deficit of non-oil-exporting developing countries rose from US$18 billion in 1973 to US$45 billion in 1975, while the outstanding public debt of eighty-six developing countries rose from US$38 billion in 1965 to US$117 billion in 1973, with the external debt servicing liabilities of those countries amounting to almost 50% of new development assistance received by them.[3] The cumulative effect of all this was that the rate of growth and development in the Third World, including most of the Commonwealth Caribbean countries, which was already low, was further reduced, with a consequent increase in unemployment and a general decline in their economic well-being.

It was these realities which led to the demand for and subsequent proclamation of the NIEO during the sixth Special Session of the UN General Assembly in April 1974, which was summoned on the initiative of President Boumedienne of Algeria, then Chairman of the Non-aligned Movement, to discuss the problem of 'Raw Materials and Development' in the wake of the crisis in the international economic system which developed towards the end of 1973.

The conceptual origins of the New International Economic Order

The major achievement of the sixth Special Session was its articulation of the demand for the NIEO and its elaboration of a conceptual framework for the realisation of such an order. The resolutions[4] on the Declaration and Programme of Action on the Establishment of a New International Economic Order which were adopted during the session not only provided arguments in support of a new economic order but outlined a number of specific measures relating to trade in raw materials and primary commodities, industrialisation, the transfer of technology and the question of sovereignty over natural resources, as well as the transfer of financial resources in support of the development efforts of the Third World. For example, the resolutions urged, among other things, the improvement of market structures in the field of raw materials and commodities of export interest to the developing countries, a substantial increase in concessional financial resources, and the adoption of measures designed to promote the industrial and agricultural output of the developing countries. In addition, the General Assembly instituted a Special Programme (and a Special Fund as part of that programme) to provide immediate relief to those developing countries most seriously affected by the economic crisis.[5]

The measures adopted by the UN General Assembly during the sixth

Special Session on the establishment of the NIEO were supplemented by the adoption at the succeeding twenty-ninth regular session of the Charter of Economic Rights and Duties of States,[6] which set out a number of important norms governing inter-state relations in the international system in the field of economic co-operation. The Charter was promoted by Mexico, with the support of the Caribbean and other Third World countries. However, as in the case of the resolution on the New International Economic Order adopted at the sixth Special Session, the Charter, although adopted by a majority of the members of the Assembly, provoked considerable controversy, since a number of developed countries disagreed with some of its provisions, particularly those dealing with the exercise of sovereignty over natural resources, and therefore entered reservations on several of its articles. On the specific question of sovereignty over natural resources, while the Third World countries declared that disputes on compensation arising from acts of nationalisation should be subject to the jurisdiction of their national courts, the developed countries insisted that they should be the subject of international arbitration.

The third major resolution on the NIEO was that adopted by the seventh Special Session of the General Assembly in September 1975 on Development and International Economic Co-operation.[7] The resolution, which sought to carry forward the initiatives taken at the sixth Special Session and the twenty-ninth regular session, embodied a number of important measures dealing with international trade, industrialisation, the transfer of technology, the reform of the international monetary and financial system, the transfer of real resources to the developing countries, food and agriculture, economic co-operation among the developing countries and a general restructuring of the economic and social sectors of the UN system to facilitate the process of development in the Third World.

At this session of the Assembly there emerged a greater degree of 'consensus' between the developed and the developing countries than existed during the sixth Special Session. But, even so, the Assembly failed to achieve complete agreement on a number of crucial issues in the field of international trade, particularly in relation to the proposal for an integrated programme on commodities of export interest to the developing countries (which was deferred for consideration at UNCTAD IV); the establishment of an official development assistance target of 0·7% of GNP for the transfer of resources to the developing countries; the establishment of a link between special drawing rights

(SDRs) and development assistance and the replacement of national currencies by SDRs as the central reserve asset in the international monetary system.

The proposals submitted by the developed countries for negotiation during the session suggested that some of those countries were not ready to abandon their reservation on the need for the establishment of a new international economic order based on a comprehensive approach to development – which was the central theme in the stand taken by the developing countries – and preferred instead a piecemeal, case-by-case approach to the economic problems confronting the international community. On the issue of commodities, for example, the developed countries argued in favour of a commodity-by-commodity approach while the Third World advocated the unequivocal adoption of an integrated programme on commodities.

Moreover, the developed countries tended to see the development of the Third World as largely dependent on their continued prosperity and therefore displayed a preference for discussing the development issue in that context rather than as a problem in its own right. Thus it was argued that international action should be aimed at preserving the prosperity of the developed countries as a basic condition for promoting growth in the Third World. This perception continues to inform the approach of some developed countries, most notably the USA, Canada, the UK, West Germany and Japan, to the negotiations on the establishment of the NIEO.

Nevertheless, the resolution on the establishment of a New International Economic Order adopted at the sixth Special Session, the Charter of Economic Rights and Duties of States and the resolution on Development and International Economic Co-operation together succeeded in providing the conceptual framework for a fundamental restructuring of international economic relations and became, as it were, the guiding principles of the developing countries in their approach to the discussion of international economic issues.

During the negotiations which took place at the sixth Special Session, the twenty-ninth regular session and the seventh Special Session, the Caribbean countries were active and interested participants. Guyana, in particular, in its capacity as the co-ordinator of the Trade, Transport and Industry sector of the Non-aligned Action Programme for Economic Co-operation, emerged as a leading spokesman on economic co-operation among developing countries during these meetings. In this role it was responsible for tabling draft proposals on the subject and negotiating the drafts both formally and informally within the Group of Seventy-seven and in the consultations with the developed countries. With the support of Caribbean and other Third World countries, Guyana succeeded in promoting in the Second Committee a comprehensive resolution on the subject during the twenty-ninth session, as it had done during the previous twenty-eighth session. Moreover, the Commonwealth Caribbean countries worked closely with other Third World countries during the sixth Special Session to ensure the adoption of a Special Programme and a Special Fund as part of that programme designed to provide assistance to the most seriously affected (MSA) countries to enable them to deal with immediate short-term problems deriving from the international economic crisis. Given the economic difficulties faced by many of these countries, this programme was seen by them as an indispensable element of the overall package of measures aimed at ensuring their economic survival. In fact Guyana, which was subsequently designated an MSA country, was allocated the sum of US$550,000 under the Special Programme, which was used for the purchase of fertilisers and medical equipment and supplies. In addition, Guyana was granted a twenty-five-year interest-free loan of approximately US$12 million by the Venezuelan government as part of that country's contribution to the UN Special Programme.

However, the conclusion of the Lomé Convention between the ACP states and the European Economic Community in January 1975 imposed on the Caribbean and other ACP countries the need to defend the principle of differential preferences among developing countries which was implicit in the special arrangements granted to these countries under the terms of the Convention. During the seventh Special Session the issue created some controversy within the Group of Seventy-seven between the Caribbean and other ACP states, on the one hand, and some Latin American members of the Group, on the other, who sought to extend to all developing countries the preferences granted to these countries.

The specific context of the debate was the discussion of the continuation and improvement of the Generalised Scheme of Preferences (GSP) through the liberalisation of its terms and the extension of its coverage. As primary producers dependent on export markets for their products, the Commonwealth Caribbean and other ACP countries had a vital interest in defending the principle of discriminatory preferences among developing countries. However, some Latin American countries, particularly Colombia and Brazil, were reluctant to concede this principle because of their interest in securing more favourable access for their own products in West European markets.

After a lengthy debate within the Group of Seventy-seven the issue was finally settled in favour of the Caribbean and other ACP countries, largely because of their numerical superiority within the Group. The formulation which was finally included in the relevant section of the resolution adopted by the seventh Special Session provided, among other things, for the extension of the Scheme beyond the ten-year period originally envisaged and for its improvement through wider coverage, deeper cuts and other measures, bearing in mind the interests of those developing countries which enjoyed special advantages and the need for finding ways and means for protecting these interests.[8] It should be mentioned in this context that the original intention was that loss of preferential access for some products should be compensated for by the grant of preferences to other products. However, many ACP countries subsequently found this alternative unattractive in practice, given their overwhelming dependence on the limited number of commodities which enjoyed preference under the Lomé Convention.

This stand by the Caribbean countries on the GSP should not be interpreted as a narrow preoccupation with their own interests in the discussions on the establishment of the NIEO. In fact the Caribbean countries were genuinely committed to a fundamental restructuring of the existing international economic system and consistently sought to advance the dialogue on the subject, as the Commonwealth Prime Ministers' Conference in Kingston, Jamaica, in May 1975 illustrated.

The Commonwealth Prime Ministers' Conference, Kingston, Jamaica

In the context of the ongoing discussions on the NIEO, the Commonwealth Prime Ministers' Conference in 1975 was important in two respects: firstly, the presentation by Guyana, on behalf of the Caribbean countries, of the case for the NIEO and, secondly, the appointment of a Commonwealth Group of Experts to identify measures to advance the establishment of the new order.

The Guyana statement sought to restate and elaborate the position of the Group of Seventy-seven adopted at the sixth and seventh Special Sessions and urged the adoption of an integrated and comprehensive programme to eliminate the disparities between the developed and developing countries instead of a piecemeal reconstruction of the existing order. As the statement emphasised, 'it will not do to deceive ourselves about the nature of the task which must be attempted. We must embark on building a new order. It is not a task involving the repair, renovation or piecemeal reconstruction of the old order.'[9] This approach stood in sharp contrast to the British presentation,[10] which to a large extent was representative of the position of the developed Commonwealth countries, and which advocated the adoption of a commodity-by-commodity approach as a means of dealing with the economic problems of the Third World.

The second major decision of the Kingston conference that was relevant to the ongoing dialogue on the new international economic order was the appointment of a Commonwealth Group of Experts to draw up a comprehensive and interrelated programme of practical measures aimed at closing the gap between the rich and poor countries. In carrying out its work the Group was asked to address itself to the issues and proposals elaborated in the Declaration and Programme of Action on the Establishment of the New International Economic Order adopted at the sixth Special Session; the relevant principles of the Commonwealth Declaration adopted in Singapore in 1971; and the concepts and proposals advanced during the discussion on the international economic situation at the Kingston meeting, including the presentations by the government of Guyana on behalf of the Caribbean and by the British government.

The Group of Experts which met under the chairmanship of Alister McIntyre, then Secretary General of CARICOM, prepared in July 1975 an interim report for consideration by the Commonwealth Finance Ministers' meeting in Georgetown, Guyana, in August 1975.[11] It contained proposals for a number of specific measures in the field of commodities; trade liberalisation and access to markets; economic co-operation among developing countries; food production and rural development; industrial co-operation and the transfer of technology; the transfer of resources; invisibles and international institutions. A further report by the Group was prepared and published in March 1976 in anticipation of UNCTAD IV and was subsequently considered by the Commonwealth Finance Ministers' meeting in Hong Kong in September 1976.[12] The final report of the Group, which embodied the material contained in the two previous reports, was published in 1977 for consideration at the Commonwealth Prime Ministers' Conference held in London in June.[13]

The report contained useful proposals for alleviating the problem of poverty and deprivation in the Third World and for establishing the basis of a new international economic order. As the report stated, 'the developed countries, especially those with balance of payments surpluses, are well placed to take a lead in supporting the efforts of developing countries by direct measures and, indirectly, through policies aimed at generating greater strength in the world economy'.[14] But, as with so many other blueprints for change in the international economic system, the lack of political agreement on the part of many developed countries proved a major obstacle to the implementation of its recommendations.

By the end of 1975 the basic conceptual framework of the NIEO had been comprehensively outlined in the resolutions adopted at the sixth and seventh Special Sessions of the General Assembly and in the Charter of Economic Rights and Duties of States adopted at the twenty-ninth regular session. The task facing the international community thereafter was to translate principles into action. Accordingly, since 1976 the focus of attention has shifted to the search for solutions to specific issues relevant to the realisation of the objectives of the NIEO. The following conferences held during this period were of particular importance both in terms of illustrating the nature of the progress – or lack of progress, as the case might be – in the negotiations on the NIEO and the nature of Caribbean participation in those negotiations. In this context special attention will be paid to the strategy articulated by Caribbean countries in seeking to effect changes in the structure of the international economic system.

UNCTAD IV

At UNCTAD IV, which was held in Nairobi in May 1976, Jamaica was elected chairman of the Group of Seventy-seven and thus played a critical role in negotiating the platform adopted by the Group at its Ministerial meeting in Manila in February 1976. This election was due largely to the prominent role played by the Jamaican permanent representative in Geneva during the preparatory meetings held prior to UNCTAD IV and to the activist policy of the Manley government in the field of international economic relations. Moreover, because of its traditionally active involvement in the negotiations of economic issues in the United Nations and other international fora, Guyana was elected Vice-chairman of Negotiating Group IV, which dealt with the problems of the least developed countries, landlocked and island developing countries. It also chaired the Drafting Group appointed by the Conference on Economic Co-operation among Developing Countries. In addition, Trinidad and Tobago was appointed spokesman of the Group of Seventy-seven on the problems of island developing countries – an agenda item in which the Caribbean countries, Trinidad and Tobago in particular, had taken a special interest in view of their insular character.

Regarding the question of economic co-operation among developing countries (ECDC), for which, as mentioned previously, the conference appointed a special drafting group, the resolution adopted on the subject urged the developed countries and institutions of the United Nations system to support the efforts of Third World countries to promote economic co-operation among themselves. The resolution also approved the proposal of the third Ministerial meeting of the Group of Seventy-seven in Manila for the establishment of a main committee of the Trade and Development Board on Economic Co-operation among Developing Countries but stipulated that the committee should be concerned solely with the examination and review of support measures of the developed countries to the programme of economic co-operation among Third World countries and not with the content of the programme, which, in the view of the developing countries, fell within their own exclusive competence. In addition the conference endorsed the proposal by Mexico for the convening of a group of experts on economic co-operation among Third World countries in Geneva in July 1976 in order to prepare for the Conference on ECDC which the Group of Seventy-seven had decided should be held in Mexico City in September, 1976.

In the discussion of this item there was a potential conflict between the Non-aligned Group and the Group of Seventy-seven, since the former was looking forward to the Non-aligned summit in August 1976, while the latter looked forward to the Conference on ECDC which was to be held in Mexico in September 1976. While there was a commitment to ECDC on the part of both groups, the Non-aligned Group was reluctant to see its own programme on ECDC, endorsed at summit level, subordinated to the more moderate proposals of the Group of Seventy-seven, adopted at the Ministerial level. As Chairman of the Drafting Group of the Conference on the item, Guyana used its

The New International Economic Order

influence to ensure the preservation of a co-operative relationship between the initiatives of the two groups by emphasising their mutually supportive character.

On the whole, UNCTAD IV assumed special significance, since it was felt that the conference provided an important opportunity to advance the dialogue on the establishment of the NIEO by adopting solutions in a number of critical areas, especially since the seventh Special Session had specifically referred the integrated programme on commodities to the conference for consideration. However, despite its impressive agenda, which covered issues such as commodities (including the integrated programme on commodities); manufactures and semi-manufactures; the multilateral trade negotiations; money and finance and the transfer of real resources for development (including decisions on the Code of Conduct on the Transfer of Technology); trade relations among countries with different economic and social systems, and a review of the institutional arrangements with UNCTAD, the conference failed to provide the expected solutions to many of the issues discussed because of major differences between the developed and developing countries.

Apart from agreement on some support measures for the manufactures and semi-manufactures of the developing countries, the main achievement of UNCTAD IV was the adoption of the integrated programme on commodities (Resolution 92 (IV)) which provided, among other things, for a number of measures designed to bolster the price of some eighteen commodities which together account for over 40% of the export earnings of the Third World. As producers of commodities such as sugar, citrus and bauxite which account for a substantial portion of their foreign exchange, the Caribbean countries had a specific interest in the programme, although it is true to say that Caribbean support for the programme also reflected solidarity with other Third World countries in Africa and Asia which potentially were the main beneficiaries of the scheme.

A key element of the integrated programme – the so-called Common Fund – was intended to finance, among other things, the holding of buffer stocks to ensure an appropriate balance between supply and demand. But while the developing countries pressed for the convening of a conference to *establish* the Fund, the Nairobi conference, at the insistence of the developed countries, merely requested the Secretary General of UNCTAD to convene a negotiating conference *on* the Fund, thereby implying that the holding of such a conference was not in itself a commitment to the establishment of such a Fund. Nevertheless, following the convening of a special negotiating conference, agreement was reached on the establishment of the Fund.

The negotiations during UNCTAD IV suggested that the bargaining position of the developed countries had considerably hardened since the seventh Special Session. They were now less willing to accede to the demands of the developing countries. This stance was perhaps not unrelated to the general economic recovery of the developed countries following the disruptions caused by the initial oil price increases.

Although they participated actively in the negotiations, Caribbean countries, like the other Third World countries, came away from UNCTAD IV with a sense of frustration and disappointment in view of the meagre results. However, they dared to look forward with guarded optimism to the Conference on International Economic Co-operation (CIEC).

The Conference on International Economic Co-operation

Jamaica, which was one of the nineteen developing countries participating in the CIEC, held in Paris, was an active participant and thus ensured a Caribbean perspective in the discussions.

CIEC was the result of efforts by the developed countries to transfer the discussion of international economic issues from the United Nations to a more limited forum in order to escape the so-called 'tyranny of the majority' exercised by the developing countries in the UN General Assembly by virtue of its universal and democratic character.

Proposed initially by President Giscard D'Estaing of France as a dialogue on energy problems between the developed countries and the oil producers, with selective representation of the oil-importing developing countries, the conference was subsequently broadened, at the insistence of the OPEC and other developing countries, to cover other subjects such as raw materials, finance and development, in addition to energy. Moreover, representation at the conference was increased in order to ensure wider participation from among the Third World as a whole. The actual representatives were decided following consultations within the Group of Seventy-seven. These consultations endorsed the participation of Jamaica, which had already been invited under the original limited formula, as one of the select representatives of the oil-importing developing countries.

The report of the preliminary meeting on preparation for the conference, which was submitted to the General Assembly during the thirtieth regular session in September 1978, provoked considerable controversy in relation to both the representation of the developing countries at the conference and the nature of the relationship that the conference should bear to the General Assembly. Some developing countries felt that the 'Paris conference', as it was sometimes called, represented a potentially dangerous tendency, since it sought to settle important international economic issues outside the framework of the United Nations. They were also concerned lest it encourage an accommodation between the developed countries and the oil producers to the detriment of the solidarity of the developing countries. This was a view held by most of the developing countries not represented at the Paris conference. A number of other developing countries felt, however, that the Paris conference represented an important initiative, since it was the first time the developed countries had agreed to negotiate seriously over controversial international economic issues, and the opportunity should not be missed. The countries of the Commonwealth Caribbean found themselves on both sides of this debate. Jamaica, for example, was in favour of the conference, while Guyana, Barbados and Trinidad and Tobago were somewhat sceptical. In the end a formula was arrived at to ensure that the decisions of the conference did not affect the primacy of the General Assembly in the discussion of international economic issues.

In the actual organisation of its work the conference (which comprised nineteen developing countries and eight developed (market economy) countries, with the European Economic Community representing its members) established four separate commissions to deal with energy, raw materials, finance and development. During the negotiations in the commissions the developed countries sought greater assurance on supplies of energy and raw materials from the developing countries, while the latter sought an improvement in their terms of trade, access to the markets of the developed countries for their exports, relief from their debt burden and an increased flow of capital and technology from the developed countries to facilitate the diversification of their internal economic structure. It is clear, too, that the OPEC countries were especially interested in obtaining technology from the developed countries to enable them to establish industries based on the exploitation of their oil resources. At the same time they sought to resist pressure from developed countries as well as other developing countries to limit increases in the price of this product without the adoption of measures to counter inflation which would reduce their oil revenues.

However, because of disagreement between the developed and the developing countries on the fundamental issues, after several negotiating sessions extending over a period of nearly two years, the conference failed to produce any significant results beyond a nominal commitment to the establishment of a Common Fund as a new entity to serve as a key instrument in attaining the agreed objectives of the Integrated Programme on Commodities, as embodied in UNCTAD Resolution 93 (IV), and a Development Fund for the least developed countries.

There was, however, considerable difference of opinion between the developed and developing countries in their assessment of the results of the conference. The developed countries regarded it as a qualified success in that it had provided an opportunity for in-depth consideration of a number of important issues, while the Caribbean and other Third World countries saw it largely as a failure, since it had not yielded solutions to the problems that were at the heart of their demand for the NIEO. For this reason the resumed thirty-first session of the UN General Assembly, convened specially to consider the results of the conference, failed to arrive at a consensus on the question and consequently concluded without taking a decision. In other words, members of the Assembly agreed to disagree.

The thirty-second session of the UN General Assembly

In the light of the inconclusive results of the Paris Conference, the General Assembly, at its thirty-second session, which was convened immediately following the closure of the resumed thirty-first session, decided that, in future, negotiations on the establishment of the NIEO should take place within the framework of the United Nations rather than in more limited forums outside it. All the Caribbean countries, including Jamaica, supported the decision.

Accordingly the Assembly decided to establish a Committee of the Whole to monitor the negotiations taking place in the various UN bodies relevant to the establishment of the NIEO. In addition the Assembly adopted a number of proposals for restructuring the social and economic sectors of the United Nations, including a proposal for the appointment of a Director General for Development and International Economic Co-operation, designed to make the system more responsive to the needs of the developing countries.

The committee subsequently held a number of sessions, but they ended in deadlock owing to the fundamental differences between the developed and developing countries regarding the extent of the committee's competence to engage in negotiations on the issues being dealt with elsewhere in the UN.

The thirty-second session of the General Assembly was of special significance in terms of Caribbean involvement in the negotiations on the establishment of the NIEO, since Jamaica was elected Chairman of the Group of Seventy-seven. As such, Jamaica was the chief spokesman of the developing countries not only on the regular economic issues which were dealt with in the Second Committee of the Assembly but also on the important question of the restructuring of the economic and social sectors of the United Nations system, which became a central issue during the session. In this capacity Jamaica played an important role in shaping policy on these issues.

The fourth Ministerial meeting of the Group of Seventy-seven, Arusha, Tanzania

Caribbean influence was also prominent during the fourth Ministerial meeting of the Group of Seventy-seven, held in Arusha in January 1979. The meeting dealt with the whole range of issues that were to be considered at UNCTAD V, namely developments in international trade, commodities, the least developed countries, landlocked and island developing countries, manufactures and semi-manufactures, technology, shipping, monetary and financial issues, economic co-operation among developing countries, evaluation of the world trade and economic situation, and institutional issues relevant to UNCTAD. Guyana, on the basis of its nomination by the Latin American group, was elected Chairman of Negotiating Group Two, which dealt with agenda item 9 (developments in international trade); item 11 (manufactures and semi-manufactures); item 13 (technology); and item 14 (shipping).

The work of the Group was particularly important for the Caribbean countries since, in the context of the discussions on manufactures and semi-manufactures, it dealt with the question of the preservation of the special preferences of the Caribbean and other ACP countries, with these countries defending their right to special preferential arrangements under the Lomé Convention, and indeed insisting on compensation for the loss of such preferences. In opposition, certain non-ACP developing countries asserted that the GSP should be extended not only in terms of its product coverage but also to embrace as many developing countries as possible. In the end, the principle of compensation was conceded for loss of preferences arising from the expansion of the GSP, largely as a result of pressure from the Caribbean and other ACP countries by virtue of their superior numbers in the Group of Seventy-seven.

UNCTAD V

Although the Caribbean countries were well represented at UNCTAD V, held in Manila in May 1979, their actual participation — with the possible exception of Jamaica — was somewhat limited. This was due largely to the spectacular failure of the conference to reach agreement on most issues and the infrequency of meetings of the Group of Seventy-seven, which have traditionally served as an important vehicle for the articulation of a Caribbean viewpoint. Nevertheless, it is important to mention UNCTAD V in the present context, since the conference served to underline the need for the Caribbean and other developing countries to devise new strategies for bringing about change in the international economic system. Not surprisingly, this became a major preoccupation of the Caribbean countries that participated in the meeting of the Co-ordinating Bureau of the non-aligned countries held in Colombo in early June, immediately following UNCTAD V, which served as an important occasion for the formulation of the economic strategy adopted at the Non-aligned summit in Havana in August 1979.

The sixth Non-aligned summit, August 1979

The sixth Non-aligned summit provided a timely and important opportunity for members of the movement to review international economic developments and to seek to formulate a common strategy to promote change in the international economic system. The conference was held against the background of increasing evidence of an *impasse* in the negotiations on the NIEO and a potential threat to the solidarity of the Third World on the question of energy.

During the discussions two distinct sets of differences emerged among members of the movement. The first was between the countries that subscribed to an orthodox Marxist-based anti-imperialism which drew a clear distinction between the developed market-economy countries and the developed socialist countries. They argued in essence that the socialist bloc countries should be seen as the natural allies of the Non-aligned countries. This view prevailed in the end, as was evident from the thrust of the documents that emerged from the conference. The other difference was that between the oil-exporting and the oil-importing countries on the question of energy and the overall strategy to be pursued by the developing countries.

Against the background of the initiatives taken at the Colombo and Georgetown meetings, Jamaica and Guyana emerged as leading spokesmen on the economic issues and as major defenders of the interests of the oil-importing developing countries. As a result of the Jamaican initiative, considerable attention was devoted to the discussion of the energy issue at the conference, mainly in informal sessions. Similarly, some developing countries, notably Algeria, Iraq and other oil-exporting countries, continued to put their faith in the continuation of the global dialogue between developed and developing countries on the traditional terms as the means of achieving the establishment of the NIEO. Contrary to this view, Guyana and other oil-importing countries urged the strengthening of collective self-reliance and mutual assistance through increased economic co-operation as a critical element in the developing countries' strategy for securing fundamental structural change in the international economic system. In the end the movement decided to close ranks and supported the call for a renewal of global negotiations between the developed and developing countries, following some modest concessions by the oil-exporting countries which endorsed the Georgetown agreement to ensure oil supplies to other Third World countries on a priority basis. No agreement was reached, however, on concessionary prices for the oil-importing countries.

A Caribbean strategy for change in the international economic system

Despite the number of international conferences that have been convened in the past decade or so to deal with the question, progress on the establishment of a new international economic order has been extremely slow. This is perhaps not surprising, since what is at stake is a fundamental restructuring of the existing international economic system in which the developed countries continue to exercise a predominant influence. It is natural, therefore, that those countries should resist changes in that system which they perceive as aimed at reducing their traditional power and influence, even though the Third World countries have argued that such changes are necessary in order to ensure prosperity to both the developed and the developing countries in the long run.

What is significant, however, is the deterioration in recent years of the bargaining position of the developing countries. In this connection it is true to say that the developing countries attained their greatest strength in the period 1973–74, following the decision of the OPEC countries to increase oil prices and the political and diplomatic support given to this measure by the developing countries as a whole, since the developed countries were fearful of the economic consequences of a massive transfer of surpluses to the developing world or, more precisely, to sections of the developing world, which this action entailed.

As it turned out, through increases in the price of manufactured goods, energy conservation strategies and the skilful operation of international monetary and financial arrangements, the developed countries, despite the persistence of low rates of growth, have been able to overcome the worst effects. Moreover, the recycling of OPEC surpluses into the economies of the developed countries has contributed to their recovery and so weakened the bargaining position of the developing countries as a whole. For this reason many have emphasised the need for the latter to reassess their strategy and identify new measures to strengthen their hand *vis-à-vis* the developed world.

Caribbean countries have been in the forefront of the search for new and appropriate strategies to effect fundamental changes in the structure of existing international relations. Both Guyana and Jamaica have taken important initiatives in this area, especially the former. Guyana has not only been appointed a Co-ordinator of the Trade, Transport and Industry sector of the Non-aligned Action Programme, in which capacity it has sought to promote a concrete programme of economic co-operation among developing countries, but from the twenty-eighth to the thirty-third session of the UN General Assembly has annually promoted, on behalf of the Group of Seventy-seven, resolutions on economic co-operation among developing countries which have been adopted by the General Assembly. Moreover, in keeping with its commitment to the principle of ECDC, Guyana hosted in August 1977 the first Ministerial meeting to promote horizontal co-operation among ACP countries and in August 1978 a meeting of representatives of producers' associations in order to promote the establishment of a Council of Producers' Associations.[15]

Apart from the concrete efforts aimed at promoting ECDC, Guyana has been a major articulator of the concept at the theoretical level. For example, at the Ministerial meeting of the Co-ordinating Bureau of the Non-aligned countries in Havana in May 1978, Guyana called on the Non-aligned movement to go beyond the conception of producers' associations and adopt a more comprehensive strategy in terms of establishing industrial complementation arrangements among the developing countries based on the utilisation of their natural resources.

Similarly, at the meeting of the Co-ordinating Bureau of the movement held in Colombo in June 1979 Guyana urged the strengthening of programmes of economic co-operation and mutual assistance among developing countries as a strategy of development based on the principle of collective self-reliance and as a means of building up a countervailing power in the Third World *vis-à-vis* the developed countries in an effort to effect changes in the existing international economic order. At the Colombo meeting, Jamaica also sought to promote a comprehensive discussion on the energy question among the Non-aligned countries. Although the decisions of the meeting on this subject were somewhat tentative, it marked an important breakthrough in the discussions on energy matters within the Non-aligned movement. In fact the initiatives taken by Guyana and Jamaica were important in setting the stage for subsequent discussions on these issues within the Non-aligned movement.

In keeping with the decision of the Colombo meeting, Guyana hosted in August 1979 a meeting of a select group of Non-aligned countries to discuss collective self-reliance and mutual assistance among the developing countries. The proposals put forward by Guyana at the meeting dealt, among other things, with a scheme for reducing the impact of oil price increases on the non-oil-producing developing countries in an effort to strengthen the solidarity of the developing countries in their negotiations with the developed ones. However, the meeting was only able to arrive at agreement on ensuring oil supplies to the developing countries on a priority basis and on promoting increased investment by oil-producing countries in the Third World.

At the subsequent sixth Summit Conference of the Non-aligned countries, held in Havana the same month, Guyana reiterated its interpretation of economic co-operation among developing countries in terms of the establishment of industrial complementation arrangements and the development of trade and transport arrangements to service this pattern of economic co-operation. It was argued that the development of such a pattern of horizontal co-operation could directly affect the basis of power in the international economic system and thus modify the present vertical dependent pattern of integration of the economies of the developing countries into the international economic system. Moreover, it was felt that unless such a pattern of horizontal co-operation was developed, the spiralling pattern of increases in oil prices to compensate for the increased price of manufactured goods from the developed countries and the depreciation in the value of the currencies of the developed world would continue, with serious consequences for the economies of the developing countries.

This involves a new perception of the strategic approach necessary to effect fundamental change in the international economic system, since it is premised on the belief that, while negotiations between the developed and developing countries continue to be important, the present international economic system cannot be changed by the mere application of logic and reason but rather by collective action on the part of Third World countries designed to build an independent basis of economic power as a means of developing the countervailing power necessary to sustain the negotiations with the developed countries in the quest for the establishment of a new international economic order.

In conclusion, it may be said that the role of the Commonwealth Caribbean states in the negotiations on the NIEO represents an interesting example of the capacity of small developing states to influence and shape contemporary diplomacy and to make a significant contribution to the dialogue on change in the international economic system. The role of Guyana and Jamaica has been specially highlighted because of their activist posture in international affairs compared with other countries such as Barbados, Bahamas and Trinidad and Tobago which traditionally have adopted a low-key stance.[16] Nevertheless, it is true to say that all Caribbean states have been committed to the principle of effecting changes in the international economic system and, in different degrees, have supported the efforts to achieve this objective. Indeed, when the history of the developing countries is written the Caribbean countries, which because of their ethnic and cultural links with Africa and Asia have served as a bridge between these continents and Latin America, will be seen as an important catalyst in promoting change in the international economic system, whether acting through such varied groupings as the Non-aligned movement, the Group of Seventy-seven, the ACP Group, the Commonwealth or through direct participation in the global system.

Of course, most Caribbean countries understand that the demand for a new international economic order is not intended to replace action at the national level to promote development, since it is generally recognised that development depends to a large extent on appropriate action at the domestic level – action, for example, to own and control natural resources; to diversify the production structure as well as trading patterns; to establish sectoral linkages within the economy; and to institute programmes of import substitution as a stimulus to domestic agricultural production. But even then it will still be necessary to ensure that the international economic environment in which these economies function is so structured as to be supportive of internal development efforts, since, in the Caribbean, small size, a narrow production base, low levels of capital accumulation and technological development and the open nature of the economies constitute major obstacles – at least in the short and medium term – to the achievement of genuine self-sustained growth. For this reason Caribbean countries are likely to continue to be active participants in the dialogue on the establishment of a new international economic order.

Notes

1. The views expressed in this chapter are those of the author and do not represent the views of UNDP or any other organisation of the UN System.
2. The terms 'Third World' and 'developing countries' are used interchangeably in the text. However, in UN circles and in the Group of Seventy-seven the latter is the preferred usage, since some countries, most notably Yugoslavia, think that the concept of the Third World tends to be somewhat restrictive. Moreover, no special distinction is made between the so-called least developed countries (LDCs), even though they have special needs, and the other developing countries, since both groups of countries have maintained a common solidarity in the negotiations with the developed countries.
3. Anthony J. Dolman (ed.), *RIO Reshaping the International Order: A Report to the Club of Rome*, New York, 1977, p. 38.
4. Resolutions 3201 (S-VI) and 3202 (S-VI) in *General Assembly Official Records: Sixth Special Session*.
5. A number of countries which were designated by the United Nations as most seriously affected received modest assistance under the programme. But unfortunately the Special Fund which was intended as a vehicle for longer-term assistance had to be wound up because of lack of financial support. Many oil-importing developing countries regarded the failure of the fund as a betrayal of their cause, since it was expected that generous

contributions would be made to the fund by developed countries as well as oil-exporting Third World countries.
6   Resolution 3281 (XXIX) in *General Assembly Official Records: Twenty-ninth Session*, Supplement No. 31.
7   Resolution 3362 (S-VII) in *General Assembly Official Records: Seventh Special Session*, Supplement No. 1.
8   See para. 8 of resolution 3362 (S-VII) in *General Assembly Official Records: Seventh Special Session*, Supplement No. 1.
9   L. F. S. Burnham, *In the Cause of Humanity*, Georgetown, 1975, p. 13.
10 See statement by the Rt Hon. Harold Wilson, Prime Minister of Great Britain, distributed at the conference.
11 *Towards a New International Economic Order (Interim Report by a Commonwealth Experts' Group)*, Georgetown, 1975.
12 *Towards a New International Economic Order (Further Report by a Commonwealth Experts' Group)*, Commonwealth Secretariat, London, 1976.
13 *Towards a New International Economic Order (A Final Report by a Commonwealth Experts' Group)*, Commonwealth Secretariat, London, 1977.
14 *Ibid.*, p. 11.
15 It should also be mentioned that Guyana chaired the Economic Committee of the Non-aligned Foreign Ministers' Conference, held in Belgrade in June 1978, which took a number of important decisions relevant to the ongoing discussions on structural change in the international economic system. In addition, the government has instituted the so-called 'Turkeyen Third World Lectures' in which prominent Third World figures are invited annually to deliver a series of lectures on a topic relevant to the discussions on the NIEO. So far, President Julius Nyerere of Tanzania, Mabbub ul Haq of the World Bank, Adebayo Adedeji, Executive Secretary of the Economic Commission for Africa, and Donald Mills, Jamaica's former Permanent Representative to the United Nations, have delivered lectures under the programme.
16 No specific reference has been made to the less developed countries of the Caribbean, since they did not play a very active role in the discussion of international economic issues during the period covered by this study.

Paul Sutton

Conclusion
Living with dependency in the Commonwealth Caribbean

In the 1960s the Commonwealth Caribbean dependency economists identified a syndrome of related characteristics which defined the problems for the region, in the words of one of them, as '... mounting disorder: growing populations, lagging incomes, increasing unemployment; widening disequality, lengthening dependence and rising discontent'.[1] Although the experience of the 1970s has confirmed this prognosis as to the facts of dependence, the form has been infinitely varied, both in the way governments have sought to challenge the situation at the three levels identified in earlier chapters and in respect of the specific characteristics of each case.

This is most readily apparent at the national level. The cases given suggest three alternative strategies at work: that of confronting dependence – Grenada; that of redefining dependence by successful management strategies – Trinidad and Tobago; and that of redefining dependence through unsuccessful management strategies – the cases of Jamaica and Guyana.

The fact that Grenada is the only Commonwealth Caribbean state to address the facts of dependence squarely disposes of an argument posed in the mid-1960s – that small size in itself is a powerful independent constraint on the policies of a state. This was most fully elaborated by Demas,[2] and although the intellectual community in the Caribbean had serious reservations as to its applicability[3] the policy-makers in the area were clearly influenced by what they saw as the limitation of size. Grenada shattered this in two ways: first by moving to full political independence under Gairy in 1974 and then by seeking social and economic independence under the Provisional Revolutionary Government since 1979. In each instance the 'knock-on' effect has been considerable. With respect to the first it set in motion the second wave of formal decolonisation in the region which has ended with the assumption of independence by St Kitts-Nevis. The writ of British colonialism now runs to only a few territories the combined population of which is under 50,000. In the second instance it reverses for the smaller territories all visions of permanent mendicancy by posing radical solutions to development problems which had hitherto been thought incompatible with the fact of US hegemony in the region. The PRG's particular contribution here has been to demonstrate the feasibility of blending policies which are inwardly socialist and outwardly nationalist to achieve what to date can be seen as a truly remarkable transformation of both the spirit of the mass of the people and the material circumstances in which they live. The international opprobrium this has attracted from some quarters is as regrettable as it is understandable – if Grenada can do it, then everyone can.

While Grenada may dispose of the limitation of size, Trinidad and Tobago demonstrates the importance of resources. When the decade of the 1970s opened the prospects for Trinidad and Tobago were much like those of other Commonwealth Caribbean states; when it closed they were radically different, for which the 'gift' of petroleum was almost uniquely responsible. Trinidad and Tobago's position is thus exceptional, a fact underlined by its elaboration of a policy open to no other Caribbean state – that of developing a high-technology industrial base. In this the prime motor of initiative and accumulation has been the state, in partnership with private capital, local and international. Politics and economics, public and private interests, have accordingly become closely intertwined, not wholly separable either in theory or in practice. This has generated much criticism within the country, not least as to whether the pragmatic approach to development which necessarily arises is in itself sufficient to ensure long-run success. For the moment, however, the problems for Trinidad and Tobago, unlike those of the rest of the Commonwealth Caribbean, fall under the rubric of rapid overdevelopment rather than lingering underdevelopment. It is a question which, understandably, the literature has not addressed[4] and an outcome which, all to predictably, has attracted more envy within the region than appreciation or even interest.

The contrast with Jamaica and Guyana on this point could not be more absolute. In so far as any Commonwealth Caribbean countries in the 1970s have attracted interest from abroad it is these two. Superficially their experience has been linked, subsumed under the label of 'socialism' and dismissed as a failure. This is a hasty conclusion which the facts simply do not bear out. The nature of 'socialism' in Jamaica and Guyana has been radically different. In the former it has constituted a serious aspiration for change by a large mass of the people whilst in the latter it has become no more than a strategy for aggrandisement by a ruling elite. On this count it is clear that the experience of Guyana should properly be understood under another label. Clive Thomas, in chapter 3, is at a loss to suggest one, though he describes a process of 'fascistisation' with the parallel establishment of an authoritarian state. This development is without precedent in the Commonwealth Caribbean, although, of course, it is by no means uncommon in Latin America, where it is most often linked to the onset of a simultaneous crisis in the domestic and international economy. There is no doubting the salience of this dimension in the Guyanese case, with its bankrupt economy, collapsing state sector and acute food shortages in Georgetown, all of which point out, with brutal precision, the consequences of tinkering with dependency rather than seriously attempting to overcome it. The example of Guyana also underlines the frailty of political order in the Commonwealth Caribbean by revealing that very little other than the determination and will of the people to resist stands in the way of the establishment of a personal dictatorship.

The appropriateness of the label 'socialist' can also be raised in respect of the recent experience in Jamaica. Many of Michael Manley's reforms were no more than a manifestation of a populist nationalism designed to ensure maximum opportunities for the middle class and enhanced opportunities for lower-class Jamaicans. That it should have aroused so much controversy is thus surprising. To a degree Manley himself cannot escape responsibility for this. On many issues there was more rhetoric than substance, with the result that he was subject to intense international pressures such that he was unable to materially advance conditions at home. Increasingly squeezed in the middle, Manley ultimately abandoned all hope of putting his programme through and simply muddled along until that became untenable as well. His government thus ended on a low to which the response of the electorate was in no way surprising. At the same time the experience was not entirely negative. Manley's own reflections upon it[5] raise many practical and theoretical questions which bear upon future attempts in Jamaica or elsewhere in the region to confront the issue of dependence seriously. Perhaps the most significant is the matter of destabilisation. Without the intransigence of the IMF and the ill-will of the United States Manley might well have survived even if he would not have prospered. Yet the fact that ultimately he did not do so is not wholly attributable to external factors but has a dimension internal to Jamaica itself. Destabilisation depends on local allies willing to be the agents of a foreign power, and in Jamaica, as in Chile, these were readily found. Overcoming dependence in this context is thus a matter of confronting not simply enemies abroad but client classes at home as well.

The very different national experiences just outlined have had considerable significance in shaping policies at the regional level. Here it is important to remember that the early 1970s were years of maximum advancement and hope for the integration movement. Informing it, in no small measure, were the arguments of technocrats and members of the political intelligentsia, a number of whom were later to hold senior positions in the CARICOM Secretariat. It seemed, at the time, as if the only future for the Commonwealth Caribbean, individually and collectively, lay in the integration movement, and from outside the region it attracted considerable interest on this basis. The most lasting and explicit memento of this is perhaps the monumental report by the World Bank mission, with its carefully qualified and guarded optimism about the future of Commonwealth Caribbean integration.[6] In this, as events were to show, it was fundamentally mistaken, its expectations based on an incorrect reading of what CARICOM actually represented. In reality, this was no more than a transposition to another level of national aspirations: a means of enhancing national capacity rather than dissolving it within a broader unit. Payne has caught the 'incomplete' character of this process with his concept of regionalisation, which he defines as 'a method of international co-operation which enables the advantages of decision-making at a regional level to be reconciled with the preservation of the institutions of the nation state'.[7] On this understanding, when the case for co-operation at the regional level is deemed to conflict with the individual development of the nation, then the latter can be expected to prevail. And this is precisely what happened in the second half of the 1970s, when Trinidad's economic success raised the question of how far CARICOM was integral to its future development, just as Guyana's and Jamaica's economic failure raised questions as to how far it acted as a constraint on theirs. Under these conditions it is not surprising that CARICOM as a whole should have made so little progress, the only achievements coming in strictly defined and narrowly functional areas. The difficulties encountered in the key areas of industry and agriculture make this very plain. In the case of the former, efforts at industrial programming have been so slight and the results to date so meagre that a charge of seriousness of intent can be levelled at all concerned. In part this is the fault of the machinery established within CARICOM itself but, more important, it is the consequence of a failure of political will in the light of what industrialisation has always promised to the region's key decision-makers — growth and transformation resolving the ills of underdevelopment, the reward for which is support at the polls. With respect to agriculture the picture is somewhat brighter, though advances have been essentially *ad hoc* solutions painting in parts of the canvas without recourse to the shape of the whole. This, admittedly, is better than nothing, though in itself, of course, it is far from satisfactory. So in the end has been the 'Group of Experts' report, drawn up to survey CARICOM and make recommendations for the future.[8] In maintaining that 'integration is the only viable option for the optimal development of all the peoples of the region' it verified the validity of existing analysis but offered nothing new in respect of how it might be achieved. With the report, as within CARICOM as a whole, going through the motions of consultation and discussion is presently regarded as activity enough.

In contrast to the fragmentation at the national level and incoherence at the regional, the Commonwealth Caribbean has maintained a unity of purpose over a range of issues at the international level which has accorded to the region a specific international personality. This is particularly apparent in its relations with Britain and the EEC and in the various sets of negotiations which aspired to realise the NIEO. Of paramount concern here, of course, have been economic questions and particularly those at the very centre of any discussion of dependency — foreign investment, aid, trade and technological reliance. Demands for changes in these areas have been framed very precisely by several Commonwealth Caribbean states and pursued by them with a relentless diplomacy. Unfortunately, for the effort expended the returns have been slight. The international economic system has not been reformed, with the consequence that the hoped-for relief at the national level has not been forthcoming. The experience as a whole has been a failure, but that does not mean it has been without positive consequences. As the 1970s opened, questions could be asked about the seriousness with which the Commonwealth Caribbean identified with the Third World; as it closed, they could not. Wherever one looked to Third World political organisations — UNCTAD, the Non-aligned movement, the sixth and seventh Special Sessions of the UN, the Paris conference, Lomé and Cancun — the Commonwealth Caribbean was active in a leading role.[9]

This is not without considerable significance for the future and is a striking confirmation, if any were needed, of the generality of the underdevelopment syndrome affecting the Commonwealth Caribbean in which the individual differences among states are perceived by those within and outside the region as matters of degree rather than kind.

The same reasoning broadly applies in respect of economic relations with the Americas (North, Central and South), where the existence of a specific sub-regional headquarters for the Economic Commission for Latin America in Port of Spain and the discussions preceding the launching of the CBI variously defined the Caribbean as a distinct unit with characteristic problems in need of specifically tailored responses. In the case of the latter it also introduced and underlined the importance of the political dimension, which has lately risen to such prominence in inter-American relations (defined as including the Commonwealth Caribbean). That it has done so in the 1970s is largely for reasons external to the Commonwealth Caribbean itself, lying in the complex redefinition of traditional US–Latin American relations which continued throughout the decade. New concerns have been introduced in this, notably 'human rights' and demands for 'collective economic security',[10] now forcefully articulated by a more assertive Latin American defence of its interests in which Venezuela, Brazil and Mexico have been especially active. These countries have also demonstrated an interest in the Caribbean at a time when the area itself, and particularly Central America, began to impinge more directly on US decision-makers, leading to an escalation of US involvement in the region from 1979 onwards.[11] All this has very much been a mixed blessing for the Commonwealth Caribbean. On the one hand it promises a serious concern with their plight; on the other an unwelcome intervention in their internal affairs, notably their political arrangements, which hitherto had not been regarded as all that important. It has also generated an element of divisiveness within the Commonwealth Caribbean at a regional level and created a bitterness at a bilateral level between several Commonwealth Caribbean states not seen for many years. How in the long run this will all be resolved is far from clear, yet resolved it must be if any progress is to be made.

The experience as outlined above suggests that in the 1970s the most formative influences on Commonwealth Caribbean states have originated at the national and international levels. This has had the effect of pulling in two ways. In one direction, primarily at the national level, the diversity of experience is most readily apparent. In 1970 all independent Commonwealth Caribbean countries could be described economically as MDCs (in a regional sense) and politically as governed by an essentially unrevised Westminster system. In 1980 this no longer applied. A number of LDCs were independent, and among the MDCs significant economic differences had opened up. Nearly everywhere the Westminster system had been called into question, in some countries more than others, and in one of them, Grenada, explicitly abolished. Policies which everywhere in 1970 were derived pragmatically and based on *ad hoc* solutions, generally by 1980 carried conspicuous and differing ideological references.

Against fragmentation at the national level can be counterposed the similarity of the experience at the international. The Commonwealth Caribbean has been defined from without as a collective unit at the same time as a collective self-identity has developed from within, the one fortuitously reinforcing the other in respect of the pursuit of tangible interests in the international system. This has established a presence for the region in international politics in 1980 which it did not possess in 1970 and which in the future it may be hoped will yield economic returns so far denied.

Given the above, it is not difficult to see why solutions at the regional level have failed and the experience been so unsatisfactory. What was required was a reconciliation of the irreconcilable: of inward national assertiveness with outward collective international identity which the region ideally should have been placed to mediate but which in the event proved impossible to achieve in the economic sphere, even if in the political it has nevertheless found temporary expression in the policy of 'ideological pluralism'. The very fact that this, in itself, is founded on a recognition of national differences as paramount serves to underline and emphasise the ultimately greater significance of developments at this level. For the Commonwealth Caribbean, we may conclude, the national revolution is not yet over, all the possibilities not yet fully developed or explored. This poses problems for the future, for clearly some states are more advanced on this path than others – and yet all must tread it together if the decisive breakthrough into development and independence is to be achieved. It will require new strategies and fresh thinking, with an emphasis on the combined as well as the uneven underdevelopment of recent years. In this politicians, technocrats and the immense resources of the Commonwealth Caribbean people at large have a part to play. Against such pressures the clock surely cannot for ever stay its hand.

Notes

1. Lloyd Best, 'Independent thought and Caribbean freedom' in N. Girvan and O. Jefferson (eds.), *Readings in the Political Economy of the Caribbean*, Kingston, 1971, p. 7.
2. W. G. Demas, *The Economics of Development in Small Countries with Special Reference to the Caribbean*, Montreal, 1965.
3. See especially V. A. Lewis (ed.), *Size, Self-determination and International Relations: The Caribbean*, Kingston, 1976.
4. A sole exception here is G. Lamb, 'Rapid capitalist development models' in D. Seers (ed.), *Dependency Theory: A Critical Reassessment*, London, 1981.
5. M. Manley, *Jamaica: Struggle in the Periphery*, London, 1982.
6. Published as S. E. Chernick, *The Commonwealth Caribbean: the Integration Experience*, Washington, DC, 1978.
7. A. J. Payne, *The Politics of the Caribbean Community 1961–1979: Regional Integration amongst New States*, Manchester, 1980, p. 286.
8. See The Group of Caribbean Experts, *The Caribbean Community in the 1980s*, Georgetown, 1981.
9. The significant omission is OPEC. Trinidad considered membership in 1963 but did not seek admission until 1972, when it was vetoed by Iraq. For a short period thereafter it remained actively associated with OPEC but in 1974 interest lapsed and no moves have been made since to revive it.
10. A useful review of developments is to be found in T. J. Farer (ed.), *The Future of the Inter-American System*, New York, 1979.
11. See, in particular, Jenny Pearce, *Under the Eagle: US Intervention in Central America and the Caribbean*, London, 1981.

Index

Adams, Sir Grantley, 13
Adams, N. A., 210
Adams, Tom, 130
Afghanistan, 116
Africa, 2, 8, 25, 207–8
African, Caribbean and Pacific (ACP) Countries, chapter eight (204–37); and New International Economic Order, 264–5, 273–4, 276
Agency for International Development (AID), 189
Agriculture, chapter six (152–73); 19, 47–8, 83–4, 85, 87, 114, 183; see also Agricultural Marketing Protocol (AMP), Agricultural Production Credit Scheme (APC), Caribbean Agricultural and Rural Development Advisory and Training Service (CARDATS), Caribbean Agricultural Research and Development Institute (CARDI), Caribbean Development Bank (CDB), Caribbean Food Corporation (CFC), Farm Improvement Credit Scheme (FIC), Guaranteed Market Scheme (GMS)
Agricultural Marketing Protocol (AMP), 152, 155–8, 162, 163, 166
Agricultural Production Credit (APC), 160–1
Alleyne, D., 57
Ambursley, F., 33

Andean Common Market, 243
Angola, 29–30, 39, 93
Antigua, 181–2, 245
Argentina, 186, 241, 242
Aruba, 117
Australia, 183
Aves Island, 245

Bahamas, The, 183, 184, 186, 215, 278
Bananas, 19, 140, 154, 181; trade with EEC, 210, 213–14, 224; see also WINBAN
Barbados, 9, 14, 89, 127, 146, 190, 192, 239, 278; agriculture, 155, 161, 168; manufacturing, 134, 137, 142–3; relationship with Grenada, 115, 119; relationship with USA, 118, 179–82, 185–7; trade with EEC, 210–11, 215, 221
Barbados Progressive League, 13
Barrow, Brensley, 64
Bauxite, 3, 4, 140, 269; and the EEC, 210, 229; in Guyana, 77, 94–5, 98; in Jamaica, 19, 20, 23, 24, 27–9, 31, 36; US trade and investment, 180–3, 191, 194
Beckford, George, 3
Beker Corporation, 52
Belize (formerly British Honduras), 127, 176, 180–1; agriculture, 160, 164; aid to, 199, 215; Guatemalan claims on, 246, 255
Benham, Frederic, 132

Bermuda, 184, 185–6
Best, Lloyd, 2–4, 8, 59
Bird, Vere, 14
Bishop, Maurice, 106–7, 114, 130, 198; attitude to Grenadian internal democracy, 120–1; leads opposition to Gairy, 110–13; *see also* People's Revolutionary Government of Grenada (PRG)
Black Power Movement, The, 45–6, 110
Blackman, Courtney, 8–9
Blaize, Herbert, 109, 111; *see also* Grenada National Party (GNP)
Bolivia, 241
Bradshaw, Robert, 14
Brazil, 186, 252, 265; relationship with Caribbean countries, 177, 238–43, 244, 247–8, 250, 253, 255
Brewster, Havelock, 3, 8, 50, 138–40, 142, 145, 146
Britain, chapter eight (204–37) and 127–8, 177, 253, 266; and Caribbean debts, 189–90; and Caribbean migration, 127, 192; relationship with Grenada, 112, 116, 118; *see also* Colonial Office
British Guiana, *see* Guyana
British Virgin Islands, 160
Bureau of the Non-Aligned Movement, 116
Burnham, Forbes, 89, 92, 97

Callaghan, James, 34
Canada, 31, 189, 195, 198, 241; and Caribbean migration, 127, 192; relationship with Grenada, 112, 119
Cardoso, F. H., 7
Caribbean Agricultural Research and Development Institute (CARDI), 161–2, 163, 167, 168
Caribbean Agricultural and Rural Development Advisory and Training Service, (CARDATS), 162, 167
Caribbean Basin Initiative (CBI), 177, 195–202, 247, 286; *see also* USA
Caribbean Community and Common Market (CARICOM), chapters five (131–51) and six (152–73) and 119, 176, 199, 219, 227, 284–5; and middle powers, 247, 255; formation and progress, 128–30
Caribbean Development Bank (CDB), 144, 176, 219, 222, 245; and regional agricultural planning, 159–61, 163, 167, 168; relationship with middle powers, 247, 254; US aid via, 118, 189, 200–1
Caribbean Food Corporation (CFC), 163, 166–9
Caribbean Free Trade Association (CARIFTA), 127–8, 135–41, 152, 155–6, 206
Caribbean Investment Corporation, 168
Carriacou, 105, 111
Carrington, Edwin, 226
Carter, Jimmy, 34, 116; *see also* USA
Castro, Fidel, 29, 36; *see also* Cuba
Central American Common Market, 183
Centre for Industrial Development (CID), 221–3, 230
Chaguaramas, Declaration of (1970), 46
Chaguaramas, Treaty of (1973), 128
Chambers, George, 64, 70–1
Charles, Eugenia, 201
Cheysson, Claude, 228
Chile, 118, 186
Clarke, Ellis, 69–70
Coard, Bernard, 110–11, 112–13, 119
Colombia, 186, 198, 252, 265
Colonial Development and Welfare Organisation, 14
Colonial Office (Britain), 15, 43, 85, 87; *see also* Britain
Committee for Industrial Co-operation, 221, 223

Index

Commonwealth Caribbean Regional Secretariat, 138
Commonwealth Group of Experts, 265–7; *see also* NIEO
Commonwealth Sugar Agreement, 205, 206, 211–12; *see also* sugar
Conference on International Economic Co-operation (CIEC), 270–2; *see also* NIEO
Costa Rica, 199, 238
Cuba, 9, 50, 93, 238, 241, 242, 251, 254–5; accord with Manley, 29–30, 36, 39; involvement in Grenada, 108, 112–13, 115–18; relationship with CARICOM, 176, 178; relationship with middle powers, 246, 248–50; US attitudes towards, 183, 193, 195

Daily Gleaner (Jamaica), 31
Danns, G. K., 99
D'Estaing, Giscard, 270
Demas, William, 135–7, 139, 209, 281
Democratic Action Congress of Trinidad and Tobago (DAC), 67
Democratic Labour Party of Trinidad and Tobago (DLP), 44, 48
Dominica, 108, 215
Dominican Republic (formerly Santo Domingo), 176, 183, 194, 242, 247
Duncan, D. K., 31, 33, 35–6

East African Common Market, 205
Eastern Caribbean Common Market (ECCM), 143–4, 209
Echevarria, L., 252
Ecuador, 186
El Salvador, 197–99
Emerson, Ralph, 46
Enders, Thomas, 117
European Development Fund (EDF), 216–22
European Economic Community (EEC), chapter eight (204–37) and 127, 128, 177, 264, 271; aid to Grenada, 117–19

European Investment Bank (EIB), 216, 221, 230
Export-Import Bank, 189–90; *see also* USA

Faletto, E., 7
Farm Improvement Credit Scheme (FIC), 160
Federation, The (1958–62), 15–16, 44, 134
Ferreira, Ferdi, 65
Foley, Maurice, 218
France, 119, 218, 270
Frank, André Gunder, 1, 5, 7
Free West Indian, 121
Furtado, C., 1

Gairy, Eric, 14, 105, 113, 176, 198, 281; deals with opposition, 110–12; leads counter-revolution, 112; rise to power, 108–9
Generalised System of Preferences (GSP), 148, 183, 196, 200, 264–5
German Democratic Republic, 116
Girvan, Norman, 3, 8, 33, 37
Gomes, Carlton, 64
Gonsalves, Ralph, 9
Grace, W. R., 52, 60
Grenada, chapter four (105–25) and 9, 176, 210, 245, 254, 281–2; excluded from CBI, 195, 201
Grenada Committee of Concerned Citizens (CCC), 110
Grenada Manual and Mental Workers' Union, 108–9
Grenada National Party (GNP), 109–12
Grenada United Labour Party (GULP), 109
Grenadian Voice, 120
Group of Seventy-seven, The, 264, 265, 268, 270, 273, 274, 276, 278; *see also* NIEO
Guaranteed Market Scheme (GMS), 152, 157–8, 163
Guatemala, 246, 255
Guyana (formerly British Guiana), chapter three (77–104) and 52,

118, 127, 128, 146, 192, 238, 240, 248, 254, 255, 282–3; agriculture, 153–5, 161, 164; demands for NIEO, 259, 263–6, 268–9, 273–8; foreign policy, 112, 176; manufacturing, 134, 137, 142, 143, 183; relationship with the EEC, 209, 210–11, 215, 225, 229; US trade with and involvement in, 180–1, 183, 185–7, 189–90, 195; Venezuelan claims on, 241, 244, 251
Guyana Chronicle, 168

Haig, Alexander, 195
Haiti, 176, 183, 242
Heads of Government Conference, 130, 176; and regional food policy, 163, 166; on economic integration, 141, 143–4; on EEC, 206
Hewitt, A., and Stevens, C., 229
Honduras, 199
Hudson-Phillips, Karl, 59, 65, 68
Hughes, Alistair, 120
Hungary, 116

India, 3
Industrial Development Corporation, 22
Inter-American Development Bank, 189–90
International Bauxite Association (IBA), 28, 29; *see also* bauxite
International Monetary Fund (IMF), 119, 283; aid to Guyana, 94, 97–8; role in Jamaican economy, 32–7; US aid via, 189, 197
Iraq, 116

Jackman, O., 226
Jagan, Cheddi, 15, 85, 92
Jamaica, chapter one (18–42) and 9, 14, 89, 108, 127, 128, 146, 192, 239, 240, 282–4; agriculture, 153, 155, 161; banking, 8, 23, 31, 33, 35; demands for NIEO, 259, 267, 270–8; foreign policy, 112, 176; industry, 52, 133, 134–7, 142, 183,

184, 191; leading families, 20, 28, 31, 33; relations with middle powers, 143, 243, 245, 252, 254–5; relations with USA, 97, 180–1, 181–2, 185–7, 189–90, 191, 195, 199; trade with EEC, 210–11, 214–15
Jamaica Bauxite Institute, 28; *see also* bauxite
Jamaica Labour Party (JLP), 22, 31, 37
Jefferson, O., 4, 8, 20
Joint Endeavour for the Welfare, Education and Liberation of the People of Grenada (JEWEL), 110
Joshua, Ebenezer, 14
Julien, K., 61

King, G., 224, 225
King, Kurleigh, 129
Kissinger, Henry, 29–30
Knight, Derek, 113

Land Lease Project (Jamaica), 26
Laski, Harold, 26
Latin American Economic System (SELA), 252
Leeward and Windward Islands, 14, 16, 137, 155, 163, 214, 245
Less Developed Countries (LDCs), 127, 146, 287; agriculture, 154, 156–8, 168; aid from CDB, 159–61; aid from EDF, 218–20, 222; and EEC, 204, 206, 213–16, 218–20, 222, 230; and USA, 181–2; industrial development, 134, 141, 143
Levitt, Kari, 2–3
Lewis, Vaughan, 193–4
Lewis, W. Arthur, 132–5, 147
Libya, 116
Lightbourne, Robert, 205
Lomé Convention, The, first (1975), 148, 200, 204, 207–27, 264–5, 273; second (1979), 227–31
London School of Economics, 26

McClean, Hector, 64–5

Index

McIntyre, Alister, 1, 8, 207, 266
Mahabir, E., 61, 64
Manley, Michael, chapter one (18–42) and 8, 97, 119, 130, 176, 198, 252, 283
Manley, Norman, 13
Marxism, 4–5, 25, 85, 115–16, 249
Mexico, 119; attitude to NIEO, 262, 268; relations with Caribbean countries, 177, 238–43, 246–7, 250, 253, 255; relations with USA, 186, 194, 195, 198
Middle Powers, see chapter nine (238–58), Brazil, Cuba, Mexico, Venezuela
Mohammed, K., 64
Montego Bay Conference (1947), 15
Montserrat, 158
More Developed Countries (MDCs), 127, 129, 287; agriculture, 157–8; aid from EEC under Lomé, 204, 206, 216–17, 219–20, 222–3, 229–30; industry, 134, 146
Movement for Assemblies of the People in Grenada (MAP), 110, 119
Moyne Commission, The, 132
Munroe, Trevor, 9

Netherlands Antilles, 183, 186, 218
New International Economic Order (NIEO), chapter ten (259–80) and 93, 106, 175, 204, 225, 285
New Jewel Movement of Grenada (NJM), 110–12, 120, 121
New World Group, 1, 3–9, 20
Nicaragua, 116, 118, 119, 195, 198

Oil, 27, 239, 275; and Trinidad, chapter two (43–76); companies, 47, 50–3, 60; impact of 1973 crisis, 260–1; middle powers and, 239, 241; petroleum, 3, 60, 63, 71, 180–1, 183, 185–6, 211, 241, 245, 249; US trade in, 180–1, 183, 185–6, 194
Oils and Fats Agreement, 152, 155–6
O'Neil Lewis, J., 226, 231

Organisation for National Reconstruction of Trinidad and Tobago (ONR), 68–9, 71
Organisation of American States (OAS), 116, 189, 193, 242, 247
Organisation of Eastern Caribbean States (OECS), 108, 118–19
Organisation of Petroleum Exporting Countries (OPEC), 24, 28, 32, 94, 241, 260, 270–1, 276
Ortega, Daniel, 116
Overseas Private Investment Corporation (OPIC), 196

Panama, 186, 238
Pan-American Health Organisation, 189
Paraguay, 241
Parris, Carl, 70
Patterson, P. J., 34, 204, 207–8, 224, 225, 228
Payne, A. J., 284
People's National Congress of Guyana (PNC), 89
People's National Movement of Trinidad and Tobago (PNM), chapter two (43–76) and 15
People's National Party of Jamaica (PNP), chapter one (18–42) and 13, 18
People's Progressive Party of Guyana (PPP), 15, 85–6, 92
People's Revolutionary Army of Grenada (PRA), 108, 112
People's Revolutionary Government of Grenada (PRG), 105–7, 281–2; formed, 112–14; issue of democracy, 119–22; US opposition to close relations with Cuba, 108, 115–19
Perez, Carlos Andres, 245, 252
Peru, 186
Petit Martinique, 105
Petroleum, see oil, OPEC
Point Lisas Industrial Estate (Trinidad), 52, 61, 63
Point Saline International Airport (Grenada), 108

Portillo, Lopez, 246, 247
Puerto Rico, 86–9, 116, 132, 198, 201

Radix, Kenneth, 110
Rainford, D., 226, 228
Ramphal, Sonny, 204, 207–8, 225
Rastafarianism, 110
Reagan, Ronald, 116, 193, 247; and CBI, 195–200
Regional Food and Nutrition Strategy, The, 152–3, 162–70
Rice, 152, 155; in Guyana, 77, 94–5, 98; see also agriculture
Rippon, Geoffrey, 205
Robinson, A. N. R., 66–7
Rodney, Walter, 99, 119
Rousseau, Patrick, 28

St. John, B., 224, 225, 228, 230
St Kitts-Nevis, 13, 108, 245, 282; trade with EEC, 210–11; US tourism in, 181–2
St Lucia, 210, 215–16
St Vincent, 9, 108, 158, 215
Seaga, Edward, 199
Shearer, H., 89, 225
Somoza, Anastasio, 198
Soviet Union, and Cuba, 116, 249; and Grenada, 108, 116; and Guyana, 94; and Jamaica, 34, 39
Spiegel, S. L., 238
Stabex, 208, 209, 215–16, 229
Standing, G., 91
Stone, Carl, 21–2, 34
Strachan, Selwyn, 119
Sugar, 2, 140, 196, 269; EEC agreement, 205, 210–13, 224, 227; exports, 154, 180–1, 210; in Guyana, 77, 90, 94–5, 98; in Jamaica, 19–20, 23, 24, 27; in Trinidad, 48; see also Commonwealth Sugar Agreement, Lomé Convention
Sunkel, O., 1
Suriname, 176, 183, 218, 248
Syria, 116

Tanzania, 5, 110
Tapia Party of Trinidad and Tobago, 8
Thomas, Clive, 1, 5–9, 138–40, 142, 145, 146, 283
Torchlight (Grenada), 120
Tourism, 3; as foreign exchange earner, 181–5; in Grenada, 114, 117; in Jamaica, 19–20, 23, 31
Trade and industry, see chapters five (131–51) and seven (179–203); *also* bauxite, oil, petroleum, rice, sugar, tourism
Trade unions, 13, 21, 49, 86–7, 200
Transnational corporations (TNCs), 129, 185; in Guyana, 77, 86–91; in Jamaica, 191; in Trinidad, 52
Trinidad and Tobago, chapter two (43–76) and 8, 13, 14, 89, 108, 127, 146, 189–90, 192, 268, 278, 282; agriculture, 153, 155, 161, 169; as CARICOM member, 129–30; banking, 49; black power movement, 110; foreign policy, 94, 176, 244–5, 250–2; manufacturing, 133, 134, 136–7, 142–3; relations with the EEC, 210–11, 215; relations with middle powers, 244–5, 250–2, 255; relations with USA, 180–1, 181–2, 185–7; trade, 159, 180–1, 183, 205, 210–11, 215, 239, 254
Trinidad Express, 71
Trudeau, Pierre, 34

United Labour Front of Trinidad and Tobago (ULF), 68
United Nations, 8, 116, 162, 178, 250; role in NIEO talks, 261–5, 267, 270–3, 276; supports industrial development programme, 141, 144
United Nations Conference on Trade and Development (UNCTAD), 8, 285; and evolution of NIEO, 260, 262, 267–70, 272–4
United People's Movement of St Vincent, 9

Index

United States of America (USA), chapter seven (179–203) and 127, 177, 210, 239, 249, 282, 283, 286; aid, 34, 160, 162; attitude to Grenadian links with Cuba, 108, 112, 115–19; destabilisation of Jamaica, 20, 29–30, 31, 34–5, 38–9; involvement in Trinidad, 61, 67; relations with middle powers, 240–3, 247; *see also* CBI, Cuba, LDCs, middle powers, oil, OPIC
US Virgin Islands, 198
University of London, 5
University of the West Indies, 1, 5, 8, 9, 32–3, 138–40
Uruguay, 241

Venezuela, 100, 119, 143; and USA, 117, 186, 195, 198; interests in the Caribbean, 176, 238–50, 253–5

Waldheim, Kurt, 112
Warren, Bill, 7
Whiteman, Unison, 110–11
Williams, Eric, chapter two (43–76) and 15, 128, 130, 143, 163, 176, 250–1
WINBAN (Windward Islands Banana Producers' Organisation), 118
Workers' Liberation League of Jamaica, 31
Workers' Party of Jamaica, 9
Working People's Alliance of Guyana, 99
World Bank, The, 28, 118, 158, 189–90, 197, 284; aid to Guyana, 94, 97–8

Yaoundé Convention, 205–6
Young, Andrew, 34, 36